Teaching, Technology, Textuality

Teaching the New English

Published in association with the English Subject Centre
Director: **Ben Knights**

Teaching the New English is an innovative series concerned with the teaching of the English degree in universities in the UK and elsewhere. The series addresses new and developing areas of the curriculum as well as more traditional areas that are reforming in new contexts. Although the series is grounded in intellectual and theoretical concepts of the curriculum, it is concerned with the practicalities of classroom teaching. The volumes will be invaluable for new and more experienced teachers alike.

Titles include:

Charles Butler (*editor*)
TEACHING CHILDREN'S FICTION

Michael Hanrahan and Deborah L. Madsen (*editors*)
TEACHING, TECHNOLOGY, TEXTUALITY
Approaches to New Media

Forthcoming titles:

Gail Ashton and Louise Sylvester (*editors*)
TEACHING CHAUCER IN THE CLASSROOM

Lisa Hopkins and Andrew Hiscock (*editors*)
TEACHING SHAKESPEARE AND EARLY MODERN DRAMATISTS

Anna Powell and Andrew Smith (*editors*)
TEACHING THE GOTHIC

Gina Wisker (*editor*)
TEACHING AFRICAN-AMERICAN WOMEN'S WRITING

Teaching the New English
Series Standing Order ISBN 1–4039–4441–5 Hardback 1–4039–4442–3 Paperback
(*outside North America only*)

You can receive future titles in this series as they are published by placing a standing order. Please contact your bookseller or, in case of difficulty, write to us at the address below with your name and address, the title of the series and the ISBN quoted above.

Customer Services Department, Macmillan Distribution Ltd, Houndmills, Basingstoke, Hampshire RG21 6XS, England

---

Teaching, Technology, Textuality

Approaches to New Media

Edited by

Michael Hanrahan
Lecturer in English and Assistant Director of Academic Technology, Bates College

and

Deborah L. Madsen
Professor of American Literature and Culture, University of Geneva

First published in 2006 by
PALGRAVE MACMILLAN
Houndmills, Basingstoke, Hampshire RG21 6XS and
175 Fifth Avenue, New York, N.Y. 10010
Companies and representatives throughout the world.

PALGRAVE MACMILLAN is the global academic imprint of the Palgrave Macmillan division of St. Martin's Press, LLC and of Palgrave Macmillan Ltd. Macmillan® is a registered trademark in the United States, United Kingdom and other countries. Palgrave is a registered trademark in the European Union and other countries.

ISBN-13: 978–1–4039–4492–4 hardback
ISBN-10: 1–4039–4492–X hardback
ISBN-13: 978–1–4039–4493–1 paperback
ISBN-10: 1–4039–4493–8 paperback

This book is printed on paper suitable for recycling and made from fully managed and sustained forest sources.

A catalogue record for this book is available from the British Library.

Library of Congress Cataloging-in-Publication Data

Teaching, technology, textuality : approaches to new media / edited by Michael Hanrahan & Deborah L. Madsen.
p. cm.—(Teaching the new English)
Includes bibliographical references and index.
ISBN 1–4039–4492–X (hard) – ISBN 1–4039–4493–8 (pbk.)
1. Mass media and technology. 2. Mass media – Study and teaching. 3. Digital media. 4. English language – Data processing. 5. Humanities – Data processing. 6. Humanities – Technological innovations. 7. Humanities – Study and teaching (Graduate) I. Hanrahan, Michael. II. Madsen, Deborah L. III. Series.

P96.T42T36 2006
302.23—dc22 2005056592

10 9 8 7 6 5 4 3 2 1
15 14 13 12 11 10 09 08 07 06

Printed and bound in Great Britain by
Antony Rowe Ltd, Chippenham and Eastbourne

Contents

List of Figures

Series Preface

One of many exciting achievements of the early years of the English Subject Centre was the agreement with Palgrave Macmillan to initiate the series "Teaching the New English." The intention of the then Director, Professor Philip Martin, was to create a series of short and accessible books which would take widely-taught curriculum fields (or, as in the case of learning technologies, approaches to the whole curriculum) and articulate the connections between scholarly knowledge and the demands of teaching.

Since its inception, "English" has been committed to what we now know by the portmanteau phrase "learning and teaching." Yet, by and large, university teachers of English—in Britain at all events—find it hard to make their tacit pedagogic knowledge conscious, or to raise it to a level where it might be critiqued, shared, or developed. In the experience of the English Subject Centre, colleagues find it relatively easy to talk about curriculum and resources, but far harder to talk about the success or failure of seminars, how to vary forms of assessment, or to make imaginative use of Virtual Learning Environments. Too often this reticence means falling back on received assumptions about student learning, about teaching, or about forms of assessment. At the same time, colleagues are often suspicious of the insights and methods arising from generic educational research. The challenge for the English group of disciplines is therefore to articulate ways in which our own subject knowledge and ways of talking might themselves refresh debates about pedagogy. The implicit invitation of this series is to take fields of knowledge and survey them through a pedagogic lens. Research and scholarship, and teaching and learning are part of the same process, not two separate domains.

"Teachers," people used to say, "are born not made." There may, after all, be some tenuous truth in this: there may be generosities of spirit (or, alternatively, drives for didactic control) laid down in earliest childhood. But why should we assume that even "born" teachers (or novelists, or nurses, or veterinary surgeons) do not need to learn the skills of the trade? Amateurishness about teaching has far more to do with university claims to status, than with evidence about how people learn. There is a craft to shaping and promoting learning. This series of books is dedicated to the development of the craft of teaching within English Studies.

Ben Knights
Teaching the New English *Series Editor*
Director, English Subject Centre
Higher Education Academy

The English Subject Centre

Founded in 2000, the English Subject Centre (which is based at Royal Holloway, University of London) is part of the subject network of the Higher Education Academy. Its purpose is to develop learning and teaching across the English disciplines in UK Higher Education. To this end it engages in research and publication (web and print), hosts events and conferences, sponsors projects, and engages in day-to-day dialogue with its subject communities.

http://www.english.heacademy.ac.uk

Notes on the Contributors

Bryan Alexander is Director for Emerging Technologies, National Institute for Technology and Liberal Education (NITLE), Middlebury College. He has developed courses using digital media to teach the eighteenth century, Gothic literature, and the experience of war. He has also created curricula for teaching about the digital world, from multimedia writing to cyberculture and media studies. He has published on copyright policies, vampires in literature, critical theory, and utopias, and is working on a book about the uncanny in cyberspace.

Andrew Booth is Director, Flexible Learning Development Unit, and Professor of Online Learning in the School of Biochemistry and Microbiology, University of Leeds.

Lisa Botshon is Associate Professor of English at the University of Maine at Augusta. Recent work includes an edited volume, *Middlebrow Moderns: Popular American Women Writers of the 1920s* (2003), and articles on institutional issues within academia.

Dorothea Fischer-Hornung, Lecturer, University of Heidelberg, specializes in African-American studies, ethnic studies, and women's studies. She is the author and editor of several books and numerous papers on African-American dance and literature, ethnic crime fiction, and Native American literature. Currently she is president of the Society for Multiethnic Studies: Europe and the Americas, and editor of *Atlantic Studies*, a new interdisciplinary journal published by Routledge, UK.

Michael Hanrahan is Assistant Director of Academic Technology Services and Lecturer in English, Bates College. He has published articles on late fourteenth-century English literature and culture as well as on the cultural phenomenon of academic plagiarism.

Wolfgang Holtkamp is Lecturer, University of Stuttgart. His research interests include contemporary American Literature, hyperfiction, American culture studies, and e-Teaching. He has recently edited a collection of essays, *Rediscovering America* (2001).

Christopher Kelty teaches anthropology and science studies at Rice University. He undertakes historical and ethnographic research on free and open source software in the US, Europe, and India; open content movements; the ethics and politics of scientific research; and the history of software and linguistics.

Stuart Lee is Head of the Learning Technologies Group at Oxford University, the main e-learning centre at Oxford. He also teaches on the English Faculty at Oxford and lectures on medieval English and hypertext and electronic publishing. His projects and publications include two books, *Digital Imaging: a Practical Handbook* and *Building an Electronic Resource Collection: a Practical Guide* (co-edited with Frances Boyle), the JTAP Project "Virtual Seminars for Teaching Literature," and an online edition of three Old English Homilies.

David Lindley is Professor of Renaissance Literature at the University of Leeds. He has published on Shakespeare—an edition of *The Tempest* in the new *Cambridge Shakespeare* (2002), *The Tempest at Stratford* (2003), and *Shakespeare and Music* (2005). He has worked on the Stuart court masque, with an edition of court masques (1995), and a number of articles. Among other publications are *Thomas Campion* (1986), *Lyric* (1985), and *The Trials of Frances Howard* (1993). He is currently editing eleven Jonson masques for the new *Cambridge Ben Jonson*.

Leon Litvack is Reader in Victorian Studies and Head of Undergraduate Teaching at Queen's University Belfast. He teaches nineteenth- and twentieth-century literature, and his current research focuses on Dickens, as well as on cultural studies and post-colonial theory. He has authored numerous books, worked on Dickens for BBC radio and televivion, and is a Trustee of the Charles Dickens Museum in London. He is currently completing *The Complete Critical Guide to Charles Dickens* for Routledge, and is working on the Clarendon edition of *Our Mutual Friend*.

Alan Liu, Professor in the English Department at the University of California, Santa Barbara, began his research career in the field of British romantic literature and art, where his first book *Wordsworth: the Sense of History* (1989) explored the relation between the imaginative experiences of literature and history. In a series of theoretical essays in the 1990s, he extended the methodological work of this book by exploring cultural criticism, the "new historicism," and postmodernism in contemporary literary studies. In 1994, when he started his well-known *Voice of the Shuttle* web site for humanities research, he began to study information culture as a way to close the circuit between his longstanding concern for the fate of historical imagination and his parallel interest in technology. What is the relation between the imaginative experience of history and that of apparently instantaneous, history-less information culture? In 2004, Liu published his *The Laws of Cool: Knowledge Work and the Culture of Information*. Also forthcoming is *Local Transcendence: Essays on Postmodern Historicism and the Database*. Liu is principal investigator of the NEH-funded Teaching with Technology project at UC Santa Barbara entitled Transcriptions: Literature and the Culture of Information, and codirector of the English Departments undergraduate specialization on Literature and the Culture of Information. He is also a member of the Board of Directors of the Electronic Literature Organization (ELO) and chair of the Technology/Software Committee of the ELO's PAD Initiative (Preservation / Archiving / Dissemination of Electronic Literature). Most recently, he has started the interdisciplinary research project titled Transliteracies: Research in the Technological, Social, and Cultural Practices of Online Reading.

Jim O'Loughlin is an Assistant Professor of English Language and Literature at the University of Northern Iowa (USA). He is the coauthor of *Daily Life in the Industrial United States, 1870–1900* (2004).

Deborah L. Madsen is Professor of American Literature and Culture at the University of Geneva, Switzerland. She has been teaching with hypertext since 1993 to students of English and American Studies at both undergraduate and postgraduate levels. She is the author of numerous articles on e-learning and the pedagogical relevance of hypertext within the contexts of critical and cultural theory. She has also published more than a dozen books on aspects of American literature and literary theory.

Oliver Pickering is Deputy Head of Special Collections, Leeds University Library, and Associate Lecturer in English. He has published widely in the field of medieval English literature, and is Editor of *The Library*.

Eric S. Rabkin, Professor of English at the University of Michigan, Ann Arbor, has been using information technology in his teaching and research since 1975 and has filled leadership roles in academic computing both intramurally and nationally. His regularly offered courses include "Technology and the Humanities." He cofounded the Genre Evolution Project in 1998, a unique, IT-mediated attempt to meld the qualitative and quantitative study of culture. He has published more than thirty books of which the most recent is *Mars: a Tour of the Human Imagination* (2005).

Jeff Rice is an Assistant Professor of English at Wayne State University. He has published in the areas of new media, hypertext, and rhetoric and composition. He is the coeditor of *New Media/New Methods: the Turn from Literacy to Electracy* (2006).

Duco van Oostrum is Senior Lecturer in English, University of Sheffield. His current research revolves around American sports culture, in particular film and literature. His publications include forthcoming monographs on African-American sports literature and film (2007), and on autobiography in American culture (2006).

Glossary

A Level: the General Certificate of Education (GCE) A level is, in the current British education system, the highest postcompulsory high school qualification, also the university entry qualification. Prerequisite to A level study is successful completion of the General Certificate of Secondary Education (GCSE).

ARPANET: (Advanced Research Projects Agency Network) A data and communications network devised by the US Department of Defense that was the forerunner of the Internet.

AS Level: an autonomous qualification equivalent to the first year of study of the two-year British GCE A level qualification.

Blog: (see weblog).

CSS: (cascading style sheets) a simple programming language that web developers commonly use to control the presentation and layout of data in web browsers.

DHTML: (dynamic HTML) the coding practice that extends HTML by incorporating javascript, CSS, DOM, etc. to achieve interactive and dynamic as opposed to static Web pages.

DOM: (Document Object Model) an application programming interface (or API) that allows software applications, commonly web browsers, to access HTML.

Flash: A popular multimedia program created by Macromedia that is commonly used to extend the functionality and interactivity of web pages.

FTP: (File Transfer Protocol) a software standard for transferring files to and from computers.

HTML: (hypertext markup language) a simplified version of SGML (standardized general markup language) designed primarily for the creation of web pages viewable in a browser.

HTTP: (Hypertext Transfer Protocol) a standard that allows information to be exchanged and conveyed on the WWW.

Hypermedia: a distributed network of hypertexts and interactive multimedia including words, pictures, and sound connected by non-linear hyperlinks.

Hypertext: a distributed network of linked texts.

IM: (or Instant Messaging, also known as Chat) A real time communication service that allows users to exchange text messages in a rapid dialogic form.

Internet: the global system of interconnected computer networks that permits the exchange of information by means of various protocols (FTP, HTTP, TCP/IP, SMTP, etc.).

Java: a platform and machine independent programming language developed by Sun Microsystems.

Java applet: a software component written in Java that typically runs in a web application.

Javascript: a programming language that is commonly used to extend the functionality of web pages.

K12: primary and secondary education in the United States (or Kindergarten through twelfth grade, the final year of high school).

Mash-ups: songs or musical compositions that have been assembled from sampled portions of existing songs or compositions.

NGO: (nongovernmental organizations) advocacy groups that have no affiliation with governments or states.

Perl: a programming language originally created for use as an administrative tool for UNIX but which has become a practically omnipresent language available on all major computing platforms.

Sampling: digitally copying a section or segment of an existing sound recording and reusing it as an element in a new composition.

SMTP: (Simple Mail Transfer Protocol) the standard for transmitting E-mails across the Internet.

TCP/IP: (Transmission Control Protocol/Internet Protocol) a pair of standards that allow machines to establish connections between each other.

Toyota/GM NUMMI: (or New United Motor Manufacturing Inc.) California-based, joint venture between Toyota and General Motors.

VLE: (Virtual Learning Environment) refers to open source or proprietary software (sometimes called courseware) that promotes the development, storage, and maintenance of online teaching materials.

Weblog: web-based writing that typically takes the form of a diary. Weblogs are usually generated from software packages that are freely distributed on the Web (e.g., Wordpress).

Wiki: a web-based application, originally conceived as a collaborative writing tool, that creates web pages to which any user can contribute (including adding new or changing existing content).

WWW: (World Wide Web). The information sphere generally accessed by web browsers. The Web is one of many services offered over the Internet.

XHTML: (extensible hypertext markup language) an application of XML that has succeeded HTML as the evolving and future language of the Web.

XML: (extensible markup language) is a simplified subset of SGML (or standard generalized markup language) that was created primarily to describe data and to permit the sharing of data across different operating systems.

Introduction: From Literacy to e-Literacy

Michael Hanrahan and Deborah Madsen

The penetration of technology into the daily life of academia has forced, and in some cases, reinforced divisions within English departments and across institutions. The divisions are sometimes generational, highlighting differences between established and initially derided new-fangled approaches to the study of English literature, and they are also sometimes territorial, reinforcing contingent views of the appropriate and inappropriate areas of inquiry. The diverse and complex reasons that produce and reproduce such divisions within English studies and that have largely contributed to the discipline's partial uptake of new technologies are explored in a positive context in the essays collected in this volume. What contributors share is a sense that humanities computing has a mission to integrate IT into literary studies in the same way that print-based media have historically been integrated to the point that their medium is invisible to and taken for granted by practitioners. This mission often involves nothing less than a reconsideration of literacy, a concept that sometimes conveniently refers to the seemingly stable practices of reading and writing. Walter Ong (1986, 23), among others, has forcefully argued that Western culture takes literacy as "unquestionably normative and normal"—not unlike the assumed innate suitability of print-based media to literary studies. New media and computer technology highlight the contingency of an inherited, interiorized view of literacy. The transformations effected by new technologies underscore the conceptual limitation of traditional literacy to describe what many of us our doing when we "read" and "write" with new media. Various literate practices attend and constitute e-literacy—reading and writing, decoding and encoding, consuming and producing, using and repurposing, and so on. The many textual practices embraced by e-literacy are variously critiqued, advocated, and described in this volume. To varying degrees, the contributors engage directly or indirectly the ongoing as well as unfolding transformations to literate culture that have attended the Internet Age.

The acquisition of high order literacy skills is one of the traditional goals of a liberal education. Graduates should not only be able to read and write proficiently but they should also be able to do so critically, sensitively, and ethically. The cultivation of information literacy in the humanities curriculum consequently not only conforms to well-established goals of liberal education, but also helps realize recently identified priorities, including promoting independent learning and preparing graduates for life-long learning. Besides realizing these mutually informing objectives, ventures into e-literacy help extend our notions of reading and writing. Eric Rabkin persuasively articulates this idea: "Humanities education must extend itself beyond sequential literacy to deal with more capacious media and with diverse and flexible expectations for production and consumption." By doing so scholar-teachers will help reposition humanities education to participate more fully and vitally in widespread cultural transformations. As Rabkin notes, our graduates will live and work in an increasingly rich and diverse world of information. They will be expected to participate in this "infosphere" as both producers and consumers, writers and readers, creators and users. The merging of these roles has already begun. According to a recent Pew report (2004), practically half (or 44 per cent) of adult users of the Internet in the United States have created and published digital content.

As the dynamics of the infosphere begin to break down the distinction between users and creators of content, legal codes governing intellectual property rights become more difficult to navigate. Chris Kelty provides an overview of the history and current state of intellectual property law and considers how its continuing evolution influences the humanities. His essay underscores that new media have not only changed the experience of reading and writing but have also had a profound impact on the circulation of information—a cultural phenomenon that is controlled as much by technology as by law. Kelty's macrocosmic view of the culturally transformative powers of technology creates a context for understanding the shifting technologies underpinning English studies and variously identified and engaged by the volume's other contributors.

Just as the technologies enabling print-based books are perceived as "natural" to English studies, so Stuart Lee argues for the close synergy between IT and the study of literature, based on a list of core literacy skills for English studies listed in *The English Benchmarking Statement*: namely, the close reading and analysis of texts; the ability to articulate knowledge; sensitivity to the contexts that shape texts, their production, and reception; and bibliographic skills. In this list, the core skills required for the study of printed texts is assumed to be the same for electronic texts. Indeed, the increasing number of primary texts archived and available through the Web is one of the compelling reasons why students rely increasingly on electronic resources for their studies and why electronic texts increasingly find their way onto syllabi as well.

Beyond the provision of primary textual sources, there is, as Alan Liu argues in his essay, a close existing relation between textuality and the information technologies that control and manipulate it. Liu likens the academy, including the humanities, to a "post-industrial business" that has been corporatized in the same way as government, the military, and the health services. We may find the analogy discomforting, and for some the most powerful strategy for resistance is to ignore the changing economic, social, and political contexts in which contemporary English studies is situated. Liu argues, however, that the best strategy for the survival of the humanities is not to ignore but to engage IT in order to imagine and promote a knowledge society that is congruent with the traditions of humanities scholarship, as opposed to that of postindustrial capitalism.

In his essay, Jim O'Loughlin presents a similar argument but one contextualized by the so-called "crisis in the humanities." The decline in humanities funding and student numbers coincides for the most part with the large-scale adoption of computers in the educational environment. This coincidence leads O'Loughlin to speculate that, in a digital age, the humanities are no longer perceived as offering the computer-based knowledge that is identified with the kind of "cultural capital" sought from institutions of higher education. The challenge of providing subject-specific contexts for the acquisition of computer literacy poses, as O'Loughlin argues, both problems and possibilities.

Looming large among the problems is the resistance of humanities scholars to the description of their role as knowledge workers. Alan Liu and Bryan Alexander explore this resistance from complementary perspectives. Many scholars and teachers dislike this designation, which cuts across the historical role of the humanities scholar as the arbiter and inculcator of "taste." Whatever cultural capital may once have resided in the acquisition of "taste," contemporary humanities scholars are involved in generating intellectual capital. The currency and sustainability of that capital is very much what is at stake.

Bryan Alexander considers the ways in which the collaborative opportunities provided by the cyberinfrastructure threaten the professional identity of English scholars and teachers. He argues that the discipline is enamoured with the romantic image of the isolated, individual scholar-reader-writer, whose work (both in the sense of activity and product) is personal. Alexander considers the ways in which this fantasy is rapidly becoming impossible to sustain, and paradoxically why it remains stubbornly in place for now.

From a systemic perspective, Alan Liu argues that a wholesale adoption of IT by the humanities would necessitate a complete restructuring of the existing professional organization according to divisions, departments, colleges, as well as committees and classroom arrangements. We must then change not only what we teach but also how we teach it and how we handle the administration of that teaching. Along the way, we must reassess to whom we give the important task of realizing this restructured mission. At present, in many institutions of higher education the teaching of what are considered to be "mere" technical skills is delegated at best to junior faculty and to low-status classrooms as opposed to the literature seminar and at worse to nonacademic support staff who remain outside of the classroom all together. In tandem with this delegation is the awareness that teaching and research in the field of "technical skills" lacks the kind of prestige that ensures promotion. Consequently, this important effort is effectively stigmatized and marginalized.

As contributors point out repeatedly in the essays herein, the new digital textuality is an appropriate subject for English studies to engage. Likewise, many share the view that the discipline and IT will ultimately converge, and the most contested site of this convergence will be the classroom. As contributors like Lisa Botshon show, the idea that new media can be taught as an extension of current practice is a serious misjudgement. Botshon, who discusses her experiences with the popular course management system Blackboard, considers the ways in which new learning technologies have been acquired by institutions of higher education with the expectation that learning will be enhanced and made more accessible by these extensions of existing classroom learning. But, as Botshon asks, what happens when the technologies fail to deliver on these expectations?

The idea that teaching and learning in the virtual classroom are somehow the same is a common misunderstanding. Botshon observes that administrators and students alike assume that distance learning and e-learning both replicate traditional classroom pedagogy. This assumption is paired with the idea, on the part of many administrators and teachers, that online courses offer a cheap and convenient substitute learning environment while maintaining largely the same pedagogy. Students suffer from an allied but different misperception: they tend to be ignorant of the prerequisite skills needed in an electronic learning environment and compound this deficiency with the thought that online courses are somehow easier than their traditional counterparts. In fact, if it is to be successful, online teaching requires a whole new pedagogy that will replicate while not remaining identical to traditional interactive classroom practices. Botshon concludes that commercial course management systems do not encourage the required innovative pedagogy, and training in the use of these systems is restricted to handling the technical aspects of the system rather than nurturing new pedagogical skills.

Botshon's experience and conclusions are reflected in those of many contributors who describe the strategies by which they have risen to the challenge of providing a new pedagogy for a digital learning environment. They have largely avoided proprietary solutions in favour of "home grown" alternatives. Stuart Lee describes the course he designed specifically to integrate IT into the English studies syllabus at Oxford University. This is an example of the kind of innovative pedagogy for which Lisa Botshon calls. Lee teaches not in a traditional classroom but in a room furnished with a suite of computers; his assessment includes not only traditional written essays but also the requirement that students construct their own websites; printed materials are used alongside electronic texts in this course, and hypertexts supplement the canon of primary texts. In contrast to Botshon, Lee was able to use Oxford's course management system as a supplement to his course rather than as the prime pedagogical vehicle.

Stuart Lee incorporates electronic texts into his course syllabus and requires that students use IT to create an electronically-produced artifact. This integration of IT into the teaching and assessment of courses would seem to be key to the success these courses enjoy, in contrast to Botshon's abortive attempt to translate a traditional classroom-based course into virtual terms. Similarly, Jeff Rice considers the transformation of the rhetoric of composition teaching into the terms of the new media in general and hypertext in particular. To the concept of hypertext as a network of linkages, Rice introduces the idea of an aesthetic and conceptual rhetoric of hypertextual composition. The aesthetics of new media—the choices, assumptions, and effects of particular rhetorical moves or strategies—provides the starting point for Eric Rabkin's discussion of the role of audience and intention in the new digital literacy.

The arguments for the relevance of English studies to the IT revolution revolve around the need for a new pedagogy as opposed to a whole new discipline of English studies. Innovative IT-based courses require an emphasis on interactive, decentred, collaborative learning that are variously described by Leon Litvack, Duco van Oostrum, Dorothea Fischer-Hornung and Wolfgang Holtkamp, and David Lindley, Oliver Pickering, and Andrew Booth. The creation of Web-based or electronically generated assessments is an important aspect of this new pedagogy.

Leon Litvack describes the course on Literature, Imperialism, and Postcolonialism offered at Queen's University, Belfast. Included among assessed components are a mandatory PowerPoint presentation and participation in the building of the cumulative web site, *The Imperial Archive*. The interactive, decentred, and collaborative nature of the teaching and learning involved in these innovative courses promotes interdisciplinary networking across geographical and temporal boundaries. Litvack reports that his students have engaged in discussions that lasted well beyond graduation, with interlocutors across the world.

Duco van Oostrum has found additional methods and modes to foster cross-cultural perspectives in a course that deliberately sets out to foster international cooperation and collaboration. At the University of Sheffield, van Oostrum team teaches a course on American Sports Literature and Film with a colleague at the University of Maine, Orono. The course relies extensively on an electronic bulletin board as the foundational teaching tool to achieve this mission.

Extending virtual interaction and collaboration beyond discussion boards, Dorothea Fischer-Hornung and Wolfgang Holtkamp, through the American Cultural Studies Onweb project, have created a global network of university teachers who conduct online courses in an international context, incorporating students from different nations into the virtual classroom. The pedagogy employed in the courses that constitute this project is described by Fischer-Hornung and Holtkamp as "constructivist": emphasizing student interaction; process-centred assessment; student-centred learning processes; collaborative working practices; learning as a social activity; and the role of the teacher as a facilitator of learning. As they stress, courses of this kind require a great deal of time on the part of the instructor and are not appropriate to a distance-learning scenario.

In a return to the text, David Lindley, Oliver Pickering, and Andrew Booth have collaborated to create a Java-based program that allows students to acquire the basic skills for reading and transcribing Secretary Hand. This exemplary project allows students to view and attempt to transcribe digitized versions of manuscripts from the special collections at the Bodington Library, University of Leeds. The program allows them to compare their transcriptions to a master transcription and thereby tutors them in the art of transcribing early modern manuscripts.

Many of the contributions rely on or advocate collaboration as an effective means to harness the transformative powers of computer technology for English studies. New communication tools beckon us to move in this direction; such a move involves crossing new or alien territories—the law, digital media, popular or commercial culture, graphic design—that create an interdisciplinary environment where English studies can flourish. The contributions in this volume map out some of that territory and provide guidance for writing and reading it.

Appendix: Chronology of communication technologies and media

c.a. 6000 BCE	Invention of first writing system (cuneiform in Sumeria)
c.a. 3100 BCE	Invention of pictographic writing
c.a. 800 BCE	Invention of the Greek alphabet
c.a. 300 CE	Invention of the codex by the Romans
1074	First European paper mill established in Jativa, Spain
1450s	Invention of movable type and the printing press (attributed to Johann Gutenberg)
1560s	Graphite pencil invented
1826	The first successful permanent photograph was produced by Joseph Nicéphore Niépce
1834	English mathematician, Charles Babbage, creates a model of his "analytical machine," a precursor to the computer
1837	Invention of telegraph (the first electronic communication medium)
1839	Louise Daguerre refines photographic process with invention of daguerreotype
c.a. 1840	Samuel Morse and Alfred Vail develop the Morse Code
c.a. 1860	Telephone invented (attributed to Antonio Meucci)
1861	First permanent, nonfading colour photograph taken by Scottish physicist James Clerk Maxwell
1870	First typewriter commercially available in Denmark (Hansen Writing Ball)
1877	Invention of the phonograph, first device for playing and recording sound (attributed to Thomas Alva Edison)
c.a. 1885	Frederick Ives invents the half-tone technique
1893	Radio (or wireless telegraphy) invented
c.a. 1895	Motion picture invented (generally attributed to Louise and Auguste Lumière as well as Thomas Alva Edison)
1902	First photograph is transmitted electronically by Arthur Korn (in Germany)
1927	First fully electronic television demonstrated by Philo Taylor Farnsworth
1941	Konrad Zuse creates his Z3, a computer prototype
1945	Vannevar Bush publishes his concept of the "memex"
1965	Ted Nelson coins the term hypertext
1969	The precursor of the Internet, the ARPANET, devised by the United States Department of Defense Advanced Research Projects Agency
1970s	First microcomputers made commercially available
1971	First word processor, Wang 1200, introduced
1977	Apple II released (first personal computer with colour graphics)
1981	IBM personal computer released
1990	World Wide Web invented by CERN researchers

Part 1

Professional, Institutional, and Cultural Contexts

1
The Humanities: a Technical Profession

Alan Liu

I have been involved for some time in academic initiatives that bring information technology into the humanities. In ways both wonderful and painful, I have learned that information technology (IT) opens an unusually direct conduit between the perspective of the academy and those of other sectors of society. I would like to harvest this experience by reflecting on what might be called the "technical" relation between the contemporary academy and society, a relation that serves as a test bed for broader speculations on the role of the academy today.

Let me begin with a supposition. Suppose that "humanities computing," "digital humanities," "technology in the humanities," "media arts and technology," and other such awkwardly-named associations and programmes will one day fulfil their mission. That mission, phrased broadly, is to integrate information technology in the work of the humanities so fully and in so entangled a manner, at once as tool, perspective, and theme, that it would seem just as redundant to add the words "computing," "digital," "media," or "technology" to "humanities" as it was previously to add "print-based." Information technology will simply *be* part of the business of the humanities along with all its other business. What then?

Then, I surmise, it will make a great deal of difference whether the incorporation of information technology in the humanities, its business, I called it, occurred with or without critical awareness of the specifically *professional* meaning of such technology in relation to other professions in which IT has a defining role. The difference I indicate, which bears on the larger situation of the academy, may be identified through a sequence of exploratory theses as follows:

1. *Humanities scholars are also knowledge workers.* Ours is the age of the "rise of the symbolic analyst" and "intellectual capital," Robert B. Reich (1992, 169–240) and Thomas A. Stewart (1997) declare, respectively, in two of the many books of popularizing economic discourse that appeared in the 1990s to dedicate the new millennium to the work of knowledge. The distinguishing feature of such knowledge work is that it is governed by an increasingly common set of institutional, disciplinary, communicational, technical, and other practical (as in the notion of "best practices") protocols for managing productive thought. Whether as tightly wrapped as an Internet transmission protocol or as fuzzy (yet nevertheless prescriptive) as "corporate culture," these protocols include all the host of standards, specifications, declarations, procedures, routines, and functions that now bind the workers of the so-called professional-technical-managerial "new class" to the postindustrial programme of efficiency-cum-flexibility.[1]

As the full title of Stewart's book (*Intellectual Capital: the New Wealth of Organizations*) indicates, the dominant protocols of knowledge work are those of business. Yet we should recognize that there are now no natural, outer bounds to business.[2] All of the following social sectors, for example, have been touched by the logic and discourse of postindustrial corporatism: the military, the health industry, government, and even nongovernmental organizations (or NGOs). Thus consider the odd conjunction between the new, logistics-driven US military with its just-in-time forces and communication networks and the antiglobalist NGOs with their own just-in-time protest forces mobilized through IT as well as "Managing Your NGO" business instruments provided by the Association for Progressive Communications (financial spreadsheets, worksheets, analysis forms, case studies, and so on).[3] To this list of institutions influenced by postindustrial business, we can add the academy, including the humanities in higher education. It is not a stretch of the imagination, after all, to see that scholars increasingly perform analytical, managerial, administrative, and other kinds of professional work that seem ripe for corporate-inspired just-in-time and total-quality reforms. The very theories of decentered meaning adopted by the postmodernist or poststructuralist humanities, Arif Dirlik (1997, 52–83) has argued, are uncannily close to those of postindustrial capitalism.[4]

2. *The professions are increasingly bound to the protocols of knowledge work specifically by information technology.* As Alexander R. Galloway (2004, 7) points out, protocol derives from Greek proto (first) + kollēma (glue): "the first leaf of a volume, a fly-leaf glued to the case and containing an account of the MS" (OED). From its first usage on, that is, protocol was an information device, a technology not just of data but metadata that anticipated what Shoshana Zuboff (1988, 9–10) calls "informating," the accretion through computerization of ever thicker and more multiple layers of information about information. I would call special attention to the "glue" in protocol, which emblematizes the essential stickiness of information technology, otherwise celebrated for its liquid, even ethereal virtuality. Precisely its liquidity, we recognize, makes IT the perfect super glue with which to coat any profession to make it adhere to the common knowledge work model. Consider, for example, the fusion of information and knowledge in the first sentences of Stewart's book (1997, ix): "Information and knowledge are the thermonuclear competitive weapons of our time. Knowledge is more valuable and more powerful than natural resources, big factories, or fat bankrolls. In industry after industry, success comes to the companies that have the best information or wield it most effectively." "Information" and the ability to "wield" it (in other words, IT) here stick to "knowledge" so closely that there is effectively no space of separation at all, no more so (in Stewart's figure) than deuterium and tritium after hydrogen fusion. TCP/IP, FTP, SMTP, HTTP, and so on—these and other IT protocols are now our ultimate glue or, staying with Stewart's metaphor, fusion elements, networking everything together in the runaway fusion explosion called the Internet.

In our specific context, this means that the protocols of knowledge work embedded in IT are one of the main vectors by which corporate assumptions now enter the academy. Copartnership, coresearch, contractor, donor, and other official relations established between major information technology firms and institutions of learning from K12 through higher and for-profit education are just the macro side of the phenomenon. The micro side consists in the way that the ordinary work of the humanities now depends largely on proprietary IT platforms and applications.[5] Just try, for example (as I have done in a letter to the editor), to get *PC Magazine* to review products from an education-industry rather than corporate perspective even on a once-a-year, single-story basis. "It's not our focus," was the succinct conclusion of the editor, Michael J. Miller, in an otherwise kind and enlightened response.[6] The fact that the majority of humanities scholars now use an application suite named "Office" to write "files" (as opposed to essays, chapters, or books) indicates the sway, subtle yet tidal, that business protocols exert. To extend the point: the collaboration features in Microsoft's *Word*, for instance, not to mention the new XML features in Microsoft's Office System 2003 that tie individually authored documents into institutional databases, tug scholarship insensibly toward the model of corporate team work that became dogma after the fabled Toyota/GM NUMMI plant in 1984. Similar features or trends occur in products from other vendors. The influence of business runs deep even in the theoretical foundations of such mainstays of contemporary computing as relational databases, whose theory emerged from the pure, abstract mathematics of set relations but whose explanatory examples (even in the publications of the legendary originator of relational database theory, E. F. Codd) often drew from business contexts (as in examples of database tables for supplies, products, customers, etc.).[7]

3. *But IT is not just functional in knowledge work; it is allegorical*. We can take a page here from Martha S. Feldman and James G. March's study, "Information in Organizations as Signal and Symbol."[8] Feldman and March (1981, 174) argue that rational choice theory alone cannot account for the enormous appetite of business for gathering and communicating excessive information that has "little decision relevance," is too late for the decision at hand, or is never considered at all. Such information dependency, Feldman and March (1981, 182–4) suggest, can best be understood through an "information behaviour" approach that views information technology as in great part a "symbolic" or "ritualistic" *performance* of rational decision-making. IT, in other words, is not just functional in the economy of knowledge work; it is also representational, a fact never more clear than during the so-called "productivity paradox" of the 1980s and early 1990s when massive business investment in IT led to no, or even declining, productivity (Figure 1.1). As I have argued in more detail elsewhere, business kept the faith in IT during these years because the true function of IT was to serve as a speculative mirror allowing the corporations to envision whole new ways of distributed, decentralized, networked, nonhierarchical, teamworked, and otherwise "restructured" work.[9]

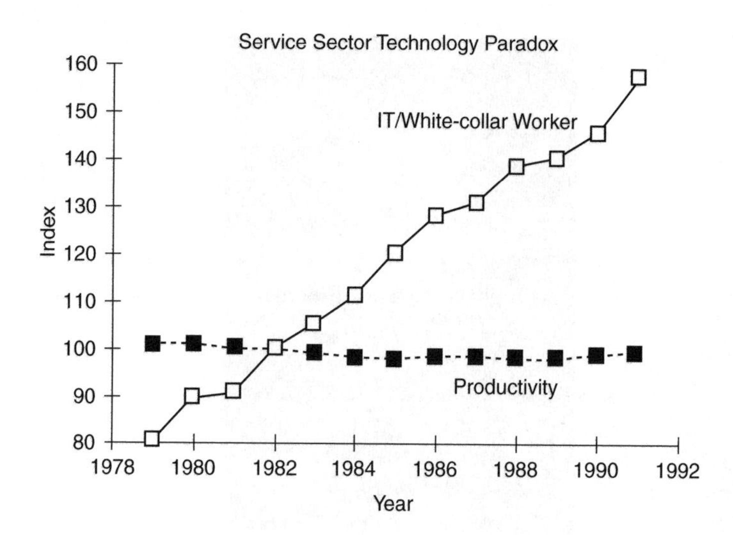

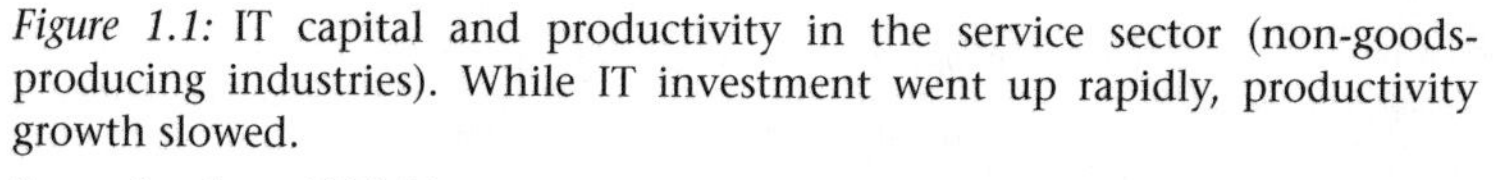

Figure 1.1: IT capital and productivity in the service sector (non-goods-producing industries). While IT investment went up rapidly, productivity growth slowed.

Source: Landauer 1995, 31.

Speculative vision, after all, has been a trope of business IT from the beginning. As Zuboff documents in her interviews, early corporate adopters of computers consistently described IT in a phenomenology of transcendental vision: IT was what let them "see it all."[10] A recent advertising campaign for IBM's middleware and information services continues the tradition. In IBM's full-page magazine ads (occurring in clustered versions several at a time on consecutive recto pages of *Business Week*, for example), workers stand like prophets with physical eyes shut but mental eyes wide open, just imagining the promised land of networked connectivity. "Can you see it?" reads the slogan (Figures 1.2, 1.3).[11] At once operational and imaginary, IT is what might be called a functional allegory or, equivalently, allegory of functionalism.[12] IT is our preeminent contemporary *poiesis*, or fictive making.

Coming now to the possible difference of humanities IT, I will close this set of theses with two in the mode of prescription.

4. *The humanities should therefore embrace the poiesis of IT for alternative ends, first of all at the level of organizational imagination.* If IT is a

Figure 1.2: "Can you see it?"
Source: *Business Week*, 17 November 2003.

Figure 1.3: "Can you see it?"
Source: *Business Week*, 17 November 2003.

poiesis, after all, we would do well to remember that humanities scholars specialize professionally in the history, forms, tropes, and, just as importantly, contradictions of poiesis, whether literary or—in the expanded, Percy Shelleyan sense—social.[13] The humanities, then, should not just adopt IT but use it in synchrony with its own traditions to imagine a society of knowledge that overlaps with, but is not necessarily the same as, that of current postindustrial capitalism. It should assert, in other words, that business has no monopoly on the use of IT for envisioning "what will be" or the "road ahead"—to cite the deterministic titles of works of IT prophecy by Michael L. Dertouzos (1998) and Bill Gates (1996), respectively. The place to start, I think, is close to home: in the alternative society that is the academy itself, where the humanities must first take care of business before it can persuasively make a case about business elsewhere.

There are two main levels on which the humanities can use IT to reimagine the protocols of the work of education. One is organizational. Business uses the functional allegory of IT to restructure. The humanities can, too—even if (and especially if) the business it needs to restructure is in crucial ways not the same as corporate business. Here I come to what I perceive to be one of the frontiers of IT in the humanities. That is the far territory on which the many, scattered humanities computing programs, centres, projects, and so on that have used IT as a catalyst to reorganize the normal disciplinary work of the humanities evolve from ad hoc organizational experiments into strategic paradigms of interest to the profession as a whole. In general, we must acknowledge, the profession of the humanities has been appallingly unimaginative in regard to the organization of its own labour, simply taking it for granted that its restructuring impulse toward "interdisciplinarity" and "collaboration" can be managed within the same old divisional, college, departmental, committee, and classroom arrangements supplemented by ad hoc interdisciplinary arrangements. The common denominator of many of these well-intentioned but institutionally insecure interdisciplinary and collaborative hacks is that they create organizational shells within which the now ingrained, individual research and teaching of the humanities can continue unchanged—with hardly any of us, for example, actually coteaching or coproducing research with anyone else in ways that exceed well-established humanities protocols (for example, multiauthor essay collections, colloquia, conferences, or panels).[14] This is despite the fact that we live in an era of declining sponsorship for individual humanities research as it has been channelled through the increasingly obsolete organizational form of the fellowship.[15] Relatively few humanities scholars thus try for large-scale project or institution based (rather than individual) funding from the government and corporations to build *structurally* interdisciplinary and collaborative programmes. And even fewer seek to initiate the systemic campus-, division-, or department-wide reorganization of the humanities that would be needed to fold interdisciplinary and collaborative work structurally into normal work (to the point, for example, of establishing course relief for grant raising and project management duties or tenurable rewards for junior faculty working on collaborative projects).[16]

Could IT in the humanities make a difference? Those in the humanities who have started funded, collaborative projects know that IT is a potential channel for refunding and reorganization.[17] There are ways of using IT to claim a place at the table where campus or external funding agencies assign monies that have worked, and many other ways that the humanities have not yet learned how to work (especially in the direction of cross-disciplinary ventures with the arts and with engineering and the sciences). One of the main tasks of those establishing programmes in humanities technology, I suggest, is to use IT to refund and reorganize humanities work with the ultimate goal *not* of instituting, as it were, Humanities, Inc., but of giving the humanities the freedom and resources to imagine humanities scholarship anew in relation both to academic and business moulds. The relation between narrow research communities and broad student audiences, for example, need not be the same as that between business producers and consumers. But unless the existing organizational paradigms for humanities work are supplemented by new models (for example, laboratory- or studio-like environments in which faculty mix with students in production work, or new research units that intermix faculty from the humanities, arts, sciences, engineering, and social sciences), it will become increasingly difficult to embed the particular knowledge of the humanities within the general economy of knowledge work. It will be difficult, for instance, to make a case before a legislature, funding agency, and ultimately the general public for the study of historical knowledges deemed obsolete by business, to analyse data through such massively inefficient methods as close reading, or otherwise to invest resources in the half-baked, buggy, never-ready-for-IPO products symptomatic of education (including student projects, dissertations, and faculty).[18]

5. *The other level on which the humanities should embrace the poetic power of IT for alternative ends is technical.* Search, query, sample, select, scan, filter, sharpen, blur, cut, paste, insert, sum, average, mark up, upload, download, attach, import, export, configure, install, save, back up, reboot, write, read (Figure 1.4). These are some of the verbs on the top-level menu of technical skills that business workers, and others participating in the common protocols of knowledge work, now need to command. By contrast, here is the usual top-level menu of the operations systematically or explicitly addressed in higher education literature classrooms (to take an example from my own native discipline): read, write, close read, contextualize, historicize,

Knowledge Work	Business	Humanities
	Search	Read
	Query	Write
	Sample	Close Read
	Select	Contextualize
	Scan	Historicize
	Filter	Interpret
	Sharpen	Critique
	Blur	
	Cut	
	Paste	
	Insert	
	Sum	
	Average	
	Mark Up	
	Upload	
	Download	
	Attach	
	Import	
	Export	
	Configure	
	Install	
	Save	
	Back Up	
	Reboot	
	Write	
	Read	

Figure 1.4: Menu of declared knowledge-work skills in business and humanities.

interpret, and critique (with the subskills required for these operations taught only unsystematically or implicitly; delegated to composition classes, lower levels of education, and IT staff; or addressed not at all). Of course, there are crucial overlaps between the two menus, especially "read" and "write." But there is also a fundamental disparity in the levels, explicitness, numbers, and granularity of technical skills.

Given the contemporary importance of technical protocols, I suggest, the time has come for the humanities to face up to its future as a technical profession like others. Only so can it give its students the necessary skills *and* (adding its normative values to those of other social sectors) impart the imagination of such skills capable of envisioning a more humane business—and, ultimately, culture—of knowledge. In short, if *technē* is where *poiesis* now lives—something that both business and the "cool" users of the newest, "bleeding-edge" technologies attest—then that is where the humanities must go.

Above all, I believe, the humanities can only teach a broader sense of culture in the age of corporate culture by demonstrating that the contemporary instinct for technical competence need not be oblivious to the sense of history that is the primary means by which the humanities at once reinforce and critique culture.[19] Technique, in other words, cannot be surrendered up to the forces of productivity as a matter of purely practical skills and competencies extrinsic to serious humanistic study. After all, theorists have been intent since at least the time of the Russian Formalists on showing that the humanities can be methodologically technical (raising the ire of those who accept the need for technical "jargon" in every single other field of contemporary knowledge except the humanities). But what this effort must be for, ultimately, is to equip humanists to reverse the field by insisting on the humanity of technique. The best way to do so is to bring to technique an awareness of historical techniques. Here are the kinds of questions to be posed in the humanities considered as a technical profession.

How might knowledge workers be educated both in contemporary information technique (the collection, verification, and collation of data; comparative and numerical analysis; synthesis and summarization; attribution of sources; use of media to produce, manipulate, and circulate results) *and* in archaic and historical knowledge technique (for example, memorization, storytelling, music, dance, weaving and other handicraft, iconography, rhetoric, close reading), with the ultimate goal of fostering a richer, more diverse, less self-centred sense of modern technical identity?

What and how did people "know," for instance, when cultures were dominated technically by orality, manuscripts, or print?

How, in other words, is the progress of knowledge constituted from broad, diverse, and always internally rifted negotiations with historical knowledges, such that every "bleeding-edge" innovation creates in its shadow not just a dark hemisphere of obsolete peoples (residual, subcultural, throwaway) consigned to the social margin, but also a repurposing and recirculation of the knowledges of the people of the margin (the true bleeding-edge)?[20]

My suggestion, to conclude, is that while the humanities must begin to teach the technical skills needed to flourish in today's society, such competence is most valuable, both to individuals and society, when wed to a full sense of the technical relationship between contemporary knowledge work and the history of human life. The humanities, a technical profession: "Can you see it?"

Notes

1. I follow Alexander R. Galloway (2004) in adopting an elastic usage of "protocol" in this essay. *Protocol* refers most precisely to the technical descriptions that standardize and regularize data formats and transmission rules allowing computers to "talk" with each other (often including both low-level and high-level formatting rules). An important example is the TCP/IP (Transmission Control Protocol/Internet Protocol) that regulates the transmission of data packets across the Internet. Depending on context, however, I also include *standards* and *specifications*, each with its own technical meaning, within a broader, more generic notion of protocols. The purpose of such elasticity of definition is to allow *protocol* to scale up in generality from its technical meaning to what Galloway (2004, 7) analyses as its formal and social or political significance, as expressed, for example, in the notion of a negotiated "diplomatic protocol"; on the formal and sociopolitical dimensions of "protocological" control, see Galloway (2004, 54–115).
2. One of the key witnesses, and/or causes, of such an outward propagation of the notion of business was the explosion of popularizing economic discourse in the 1980s and 1990s via business journalism as well as the new genre of the "business bestseller." Micklethwait and Wooldridge (1996)

survey and provide an analysis of the latter phenomenon. As I describe in *Laws of Cool* (2004, 77–8 and *passim*), this is the period when the values of "production culture" increasingly colonized consumer culture so that an ever larger proportion, both literally and symbolically, of private life began to simulate working life.

3. *Just in time* is a catch phrase in postindustrial business that originally referred to new ways of managing inventories so that parts and supplies flowed adaptively to the point of manufacturing "just in time." The phrase has since widened in usage to refer to other processes and trends of business that depend on quick-response information technology.

   The Association for Progressive Communications (APC), which supports NGOs dedicated to progressive causes, provides resources on its "Managing Your NGO" page in the categories of "Business Planning for NGOs," "Financial Management for NGOs," "Business Administration," "Marketing Strategies for NGOs," and "NGO Business Case Studies." In general, the page says, NGOs now "seek to balance sustainable business practice with their missions." "Managing Your NGO," retrieved 8 August 2004, (http://www.apc.org/english/ngos/business/index.htm).
4. The "corporatization of the university" has been much discussed of late. Critics of the corporatization of the university have included Bill Readings (1996); J. Hillis Miller (1999), published on the recto pages in a volume that contains on the verso Manuel Asensi (1999); Paul Lauter (1991, 175–97); Christopher Newfield (2003); Wesley Shumar (1997); Jeffrey Williams (1999, 742–51); and Kevin Robins and Frank Webster (1999, 168–218). For a more extensive bibliography of both scholarly and journalistic works dealing with the topic, see the "Academe and Business" section of the "Suggested Readings," on my *Palinurus* web site.
5. In recent years, back end servers at universities have increasingly shifted to open source operating systems and software. However, it will be some time, if ever, before the front end software of most academic users (that is, the programmes on actual desktops and laptops) are nonproprietary. The situation is exaggerated among humanities users, whose ordinary technology work (for example, word processing, browsing, or presenting) occurs almost wholly within proprietary standalone or client programmes more or less removed from the networking, distributed authorship, or programmer communities where open source software has made headway.
6. E-mail to the author from Michael Miller, 11 June 1998. My suggestion took the form of the following e-mail of 10 June 1998 to *PC Magazine*:

   > I'm an English professor who has subscribed to your excellent magazine for years. Your magazine, understandably, is pitched primarily at the corporate sector—meaning both the business use of computers and home use of computers by those who are in the "knowledge work" trades. How about a spread once in a while on the state of the art, and the controversies, in educational computing (both K-12 and university level)? In the context of your journal, such a story would focus primarily on available products—from general-purpose suites and OS's [operating systems]

(packaged and marketed for education) to specific application genres. For example, what products—whether business apps [applications] or education-specific apps—are on the horizon to facilitate collaborative discussion, writing, revision, version-tracking, etc., in an educational environment? (Such an environment, of course, being one that often has multiple platforms of varying levels, insecure control over its network, a varied and not wholly controllable user base, etc.) Or you could take a more strategic look at the issue—to wit: how is the educational computing market positioned right now relative to other computing markets?

7. The first example in Codd's *The Relational Model for Database Management* (1990, 8–9), for instance, refers to manufacturing parts.
8. For further citations and discussion of the symbolic (or, as I analyse it, allegorical) approach to information, see my *Laws of Cool* (2004, 153–5).
9. On the information technology productivity paradox, see Thomas K. Landauer (1995); and Paul A. Strassmann (1985). For further sources on the productivity paradox as well as discussion of its symbolic or allegorical implications, see Liu (2004, 152–4).
10. See, for example, the remarks of one of the business persons that Zuboff interviewed (1988, 163): "We'll be able to see what's happening. Not only will we have numbers, but we'll be able to see the dynamics for yesterday, today, and tomorrow. Using the projection capability, you can see immediately the impact on earnings or the portfolio. We'll be able to see the business through the terminal." For a discussion of "vision" in Zuboff with further examples, see Liu (2004, 108–11).
11. See, for example, the multipage instance of the IBM "Can You See It?" campaign in *Business Week*, 17 November 2003: 107, 109, 111. The ads shown in Figures 1.2 and 1.3 are on 107 and 109.
12. The term *allegory* may be preferred to Feldman and March's *symbolism* (1981) because we are dealing not with the iconic fusion of IT and knowledge work but instead with a contingent relation between IT as an emergent "mode of development" and knowledge work as our currently dominant "mode of production" (Castells 1996, 16–18). As in the influential de Manian analysis, allegory implies not deep fusion or integration but a shallow, congenital slipperiness or contingency a mask on the face (de Man 1983, 1984a, 1984b). IT may "stick" to contemporary knowledge-work production, that is, but not with the necessitarian telos heard in the titles of such books of information-technology prophecy as Michael L. Dertouzos, *What Will Be: How the New World of Information Will Change Our Lives* (1998) or Bill Gates, with Nathan Myhrvold and Peter Rinearson, *The Road Ahead* (1996). Rather, the representational agency of IT makes it oxymoronically sticky-and-slippery. IT as allegory harbours the imagination not just of optimal knowledge for present conditions but potentially also of *other* kinds of interfaces or masks, other kinds of knowledge, other kinds of work, even other kinds of life. Such is the semiautonomous "culture of

information," as I have argued in my *Laws of Cool* (2004), that results in the current mask of information technology as "cool." Cool people know that IT (and technology generally) serves the master of production; but they also imagine that it can represent, if only virtually, freedom—like using a workstation at the office, paradoxically, for the massively unproductive purpose of browsing cool web sites, playing online games, etc.

13. I allude to the well-known universalism of poetry in Percy Bysshe Shelley's, "A Defence of Poetry" (1821), whose last lines identify poets as "the unacknowledged legislators of the world."
14. Most often, humanities multiauthor essay collections, colloquia, conferences, and panels are fora for copresentation rather than coproduction. What is missing in the humanities is the equivalent (with suitable differences) of a "lab" environment requiring faculty to work with each other and with students in a common, goal-directed project.
15. One can be grateful for the humanities fellowships that *do* exist, even if many have been whittled away through inability to keep up with inflation, and still wish for a richer (in every sense of that word) mix of funding opportunities. Nor is it just the relative paucity of the kinds and amounts of humanities funding (by comparison, for example, with National Science Foundation grants) that is at issue. Increasingly, there is a philosophical and sociological hollowness about the basic concept of the humanities fellowship. What kind of "fellowship" is it that has the effect of emancipating the designated fellow from all (except token) demands of professional community so as to write in lonely splendour? There is a genuine need for such productive retreat as part of a broader mix of research support (including support that requires full-bore collaboration rather than retreat), but none that any longer seems to bear a relationship to the historical notion of *fellowship* The underlying rationale of the fellowship likely needs to be restated.
16. Since I originally wrote this essay, Cathy N. Davidson and David Theo Goldberg have published their important "A Manifesto for the Humanities in a Technological Age," *Chronicle of Higher Education (Chronicle Review* section) 50, no. 23 (13 February 2004): B7, retrieved 4 August 2004, http://chronicle.com (also available on the web site of the University of California Humanities Research Institute at http://www.uchri.org/humanities_manifesto.htm). Davidson and Goldberg make a point similar to mine here:

    Although humanists, for example, often engage in multiauthor, multidisciplinary projects (such as collaborative histories, anthologies, and encyclopedias) with the potential to change fields, universities and their faculties have been slow to conceive of new institutional structures and reward systems (tenure, promotion, etc.) for those who favor interdisciplinary or collaborative work. We believe that a new configuration in the humanities must be championed to ensure their centrality to all intellectual enterprises in the university and, more generally, to understanding the human condition and thereby improving it; and

> that those intellectual changes must be supported by new institutional structures and values.

Also relevant is Davidson and Goldberg's "institutional point":

> that new humanities require new structures. As we think through the revolution in electronic communication, we need to create new models for researchers to work across disciplinary boundaries, making use of databases and resources that no one scholar, or department, can maintain. That requires planning at an institutional level. We need, too, to stop talking around the issue of the single-author monograph as the benchmark for excellence, and to confront what new kinds of collaboration mean for tenure review, accreditation, and more.

Davidson and Goldberg's manifesto coincides in general direction and several particular points with my view of the relation between the humanities and technology. The one significant issue upon which I differ, as will be clear below, concerns such observations as follows in Davidson and Goldberg:

> If all we want is expertise, industry is a far better place to learn science and technology than a university. But, in fact, industry, more than anyplace else, wants not only highly trained scientists; it wants scientists who can also understand applications, intellectual property, issues of equity, human awareness, perspective, and other forms of critical analysis and logical thinking that are specifically the contribution of humanistic inquiry. The university that loses its foundation in the humanities loses, in effect, its most important asset in making the argument that "education" and not "vocational training" is worth the support of taxpayers, foundations, and private donors.

The basis of my own view of the humanities as a "technical profession" is that we are well past the era when such a clean, binary distinction can be made between the humanities and industry. (Indeed, I will argue for something like an education in the humanities *through* vocational training.) Even as the humanities have become increasingly technical, industry in its postindustrial personality as knowledge work has reciprocated by becoming increasingly humanistic. The contemporary difference of the humanities, then, cannot be understood unless we first acknowledge commonality in first principles with the new industry. That commonality sets the horizon within which the operative difference of the humanities at the present time can be discerned.

17. I cite the close-to-home example of the NEH-funded, collaborative research and pedagogy project I started with several colleagues at the University of California, Santa Barbara, in 1998 called *Transcriptions: Literary History and the Culture of Information*, which later spun off an undergraduate specialization for English majors titled "Literature and the Culture of Information" (LCI) and has collaborated with several other IT-related programmes on the UCSB campus, including the UCSB Art

Department, Media Arts and Technology Programme, and Film Studies Department and the University of California Digital Cultures Project. See the Transcriptions home page, http://transcriptions.english.ucsb.edu, and the LCI home page, http://transcriptions.english.ucsb.edu/curriculum/lci/index.as

18. Almost exactly ten years after I started it, *Voice of the Shuttle* was attacked on 31 January 2004, by a hacker through a method that exploited an underlying fault in database systems that are presented online. The attack was severe and technically challenging enough to merit "freezing" *VoS* (once it had been restored from backup) in an unchangeable state for three months. The difference between those several months and the several hours at most that a commercial operation would have tolerated is a measure of the difference between a corporate and educational enterprise. The resources and skills that education can throw at developing and maintaining its "products" is necessarily more uneven than those afforded by corporations, since faculty and students have many other simultaneous priorities (such as teaching or writing dissertations) and are never supported by a per capita level of technical staff comparable to that of corporations.
19. See Davidson and Goldberg's point that "history matters" in their "Manifesto for the Humanities in a Technological Age" (2004).
20. I borrow the concepts of the *residual* from Raymond Williams (1977, 121–7); *subcultural* from the tradition of the Birmingham Centre for Contemporary Cultural Studies (for example, Hebdige 1979); and *throwaways* from Watkins (1993).

2

A Threat to Professional Identity? The Resistance to Computer-Mediated Teaching

Bryan Alexander

Although its minimal technological underpinnings were tested and growing by the early 1970s, the Internet really became a global textual phenomenon in the 1980s. The combined corpus of e-mail exchanges and Usenet conversations involved millions of people in multiple nations. By the 1990s and the explosive growth of easy digital publishing through the World Wide Web, hundreds of millions of human beings had developed a combined textual universe so large that it cannot be effectively searched.[1] Cyberspace is a textual artifact of immense size, developed at an historically unprecedented pace, and including a rich variety of audiences, authors, discourses, and narrative production. Considered in the abstract, such an object would surely merit the attention of those who research and teach texts. Moreover, given the social and linguistic effects attending this eruption of media, critical inquiry should merit some urgency. Yet the historical record of literary scholars shows otherwise, revealing instead the construction of a marginal status for cyberspatial study and practice. A casual glance at the sample of courses listed by the Resource Center for Cyberculture Studies at the University of Washington shows only 19 out of 170 courses offered in higher education listed under English.[2]

This paper examines that resistance to computer-mediated scholarship and instruction within the discipline of English. While computer-mediated communication (CMC) has developed within a generation to the point of supporting a large, global, and accessible "cyberinfrastructure" for multidisciplinary teaching and learning (the term is the National Science Foundations'), the teaching and scholarship of writing and literature has generally been slow to take advantage of what is now the world's largest experiment in collaborative reading and writing.[3] Our discipline's resistance to cyberspace stems largely from political concerns, realized within a praxis constituted by classroom and campus resource dynamics, and supported by professional identity. Additionally, recent pedagogical and critical opportunities afforded by collaborative applications and the social software movement paradoxically strengthen resistance.

It is important not to think of this in market terms, as an opportunity for imperial growth and advantage-taking by a discipline eager to expand numbers. What is significant here is the apposite connection in terms of what each had (and has) to offer the other intellectually, and what such an intersection could offer both academia and cyberspace. An additional caveat is that the Freudian reference of this essay's title not be read to suggest an argument whereby resistance to cyberculture becomes a confirmation of the power of information technology. But we may infer in a slightly de Manian sense that English can offer a resistance to cyberculture in a productive, critical way; put another way, English should begin to read the new information technologies.

It is a cliché and at best a pseudo-problem to deem English studies scholars to be Luddites, or otherwise technology-bashers. The large body of MLA members depends heavily on modern society's elaborate network of devices and innovations, including frequent air travel, Web browsers, reliable automobiles, low- and high-cost book publishing, and so on. They have yet to adopt neoprimitivist lifestyles in response. For every faculty member who casually mentions a dislike of technology, we should be able to find twenty who regularly use electronic telecommunication tools, buy books on Amazon, and fly to multiple conferences each year.

Instead, the politics of technology is better grasped as inflected by a literary politics of personal identity. English studies has been a site of celebrating the personal since the profession began redefining itself in the 1960s. Indeed, a focus on authorial biography or textual centrality dates still further back, and persists through the present, with even the most critical theory-minded being able to self-describe as a Shelley scholar or Janeite. But we have also grown to celebrate ourselves, beyond Whitman, as we reinvested our pedagogy and critical practice with the personal. From the achievements of feminist theory and practice to the rise of standpoint theory, the politics of English studies has become increasingly personal. We presence the stories of our lives in our work in mutually enriching ways, exposing the autobiographies which shape our professional productions. We teach our students to become aware of their voices, writing for individuation. It is the firm belief of this author that this historical development is progressive. It is also synthetic, combining the constructed objects of study with how we go about constructing them and ourselves.

The popularity of many theories of fragmented, decentred, and socially constructed selves has not ultimately weakened the formation of the solitary, foregrounded, professional self.[4] For example, of all branches of scholarly inquiry, the humanities remain the most focused on sole-authored work, over against the fruits of collaboration more popular and rewarded in the natural sciences. Literature has become something like copyright law, in assuming a single creative ego behind each act of intellectual property, despite genres and cultures of influence, mixing, derivation, and collage.[5] Indeed, as it battles the digital world in ever-escalating conflicts, copyright is perhaps the most successful instance of Romanticism persisting in our contemporary world. English as a discipline, not unlike copyright, and despite evidence to the contrary, significantly prefers a cultural model of the self that is presented, central, organizing, and determining.

Cyberspace would seem to be a useful platform for such a politics of self-presentation. The Web is, after all, the home of millions of personal home pages, culminating recently in the rise of personal blogs. Web services offer a panoply of personalization, from the popular MyYahoo user portal to the self-labelling metadata sharing schemes of Flickr and del.icio.us. Yet academic and cyber forms of self-presentation collide more often than combine.

The most prominent area of collision is in the classroom experience. Computer-mediated teaching and learning has long offered the ideal of decentring the instructor within her classroom, as in the old saw of shifting the teacherly role from "a sage on the stage to guide on the side." The technologies which have appeared most prominently in cyberspace have been democratic in a formal sense, deemphasizing the professor's nominal authority and equalizing the discourse position of all members of the classroom. A professor's e-mail bears no sign differentiating it from a classmate's, nor does an instant message, chat room handle, cell phone text message, 3-dimensional space avatar, or web site URL.

The machines used in that classroom further the decentring of the English instructor by affording numerous opportunities for multitasking. Such behaviour is not necessarily a form of distraction. A student may be taking notes in Word, consulting a poem or presentation on literary history, Googling a term of poetics, posting a question on the discussion topic to a forum or blog, instant messaging a project team member, or some combination of these. Of course, the student could also be IMing a friend, surfing a sports site, or playing EverQuest. At any rate, any such activity can appear as a distraction to a teacher, or a sign of inattention. Computers do not offer the first such opportunities in the history of teaching, but their use is quite visible. Even a wholly positive multitasking (from the instructor's point of view) can threaten pedagogical centrality.

That centrality can slide into a cynosure of embarrassment all too easily, as anyone knows who has experienced technological failure in front of an audience. In business, scientific, social, or technological settings such exposed errors have certain effects and countermeasures. Yet in the literature classroom, where technology is not part of the routine, and where professors are established strongly in terms of their identity, embarrassment strikes at the heart of the role. The pedagogue arrives in the classroom with a string of credentialling behind her, coupled with a lack of technological failure expertise—the Norton Anthology has yet to display a blue screen of death, and replacing a pencil is not a resource expenditure issue. A guide on the side has less at risk in such cases than a sage on the stage. And their students maintain a comfortable lead in technological fluency over their teachers (Levin 2002). Additionally, cyberinfrastructure-based pedagogies appear to be external to English subject matter and disciplinary tradition in the classroom, which compromises attempts to construct an integral English studies sense for class and faculty. While some critics have done innovative work in cyberculture studies, this remains a very small subfield, far from the discipline's centre, and not part of the general English pedagogical mission, nor its professional identity. While some pioneers have developed English-specific technologies, such as Daedalus, the overwhelming majority is generic to cyberspace. The technology in the classroom is something other than English in a disciplinary sense, yet its often inherently textual nature is intimately woven into the practice of English studies.

One backformation in response to these classroom intrusions is to romanticize the preInternet classroom as a space where rapt students gaze upon a charismatic lecturer, mediated only by books and papers. That fantasy aside, an English instructor's sense of self can be unsettled by the technology at an explicitly political level. The destabilizing dynamic reverses direction beyond the classroom once the English faculty member's use of technology requires campus resources for support. The leading courseware products today (WebCT and Blackboard) usually drive little university friction in their use, as they are focused applications ideologically and practically, and are increasingly campus standards. But creating digital documents beyond basic web pages or courseware documents thrusts instructors into a political problematic, one requiring resources at the divisional or campus level. For example, significant amounts of web traffic, especially for rich media files, can create unusually high server demands, requiring an IT response. Or in recapitulation of the public intellectual role, a widely-read professor can be read critically against a perceived campus mission by influential figures; this is a new twist on an old issue, but heightened by the Internet's facilitation of a global audience.

The question of drawing on IT resources is a telling one for English. Unlike scientists, humanists are generally socialized to act individually, not within collaborations, and therefore often lack the professional support for approaching instructional technologists, librarians, and other faculty, especially in times of economic contraction. Moreover, graduate school experience rarely includes interaction with IT staff as part of the curriculum. The very recent nature of CMC-based campus collaborations increases the hurdle to be overcome in simply learning how they function, not to mention selecting software and hardware, iterative development, creating assessment criteria, and the very thorny matters of maintenance and archiving. In short, the political capital to be expended on campus is considerable, the risks large, and the breaking out of English's traditional role considerable.

Beyond campus cyberculture offers a different, yet also problematic dynamic for English studies. English professional identity does not prominently feature either teaching with or the critical study of digital technologies. Both aspects of the Internet remain marginal, despite years of history and enormous amounts of work by some in English. Observing this phenomenon clearly suggests a strategic environment for those in graduate school, or contemplating publication topics. Beyond such self-reinforcement, another factor is the relative importance of literature study over composition within English studies in American universities. Composition has historically tended to approach technology more energetically than has literary studies. Writing instructors identified and worked with technologies like Multi-Object Orientations (MOOs) starting more than a decade ago, seeking to take advantage of the new writing environments. A group at the University of Texas English department produced the Daedalus collaborative writing environment in the late 1980s, and have seen it through numerous technological transformations to the present day. The journal *Computers and Composition* has existed without a literature peer since the iconic year of 1984. More recently composition instructors have pounced on newer technologies, such as blogs and wikis.[6] Obviously composition studies had an opportunity to work with tools so clearly in its purview. Yet literature also identified a similar opportunity, as seen in the works of scholars like Michael Joyce, Jay Bolter, N. Katherine Hayles, Stuart Moulthrop, and Janet Murray. Hypertext, then cybertexts, and digital gaming offered (and continue to offer) a rich field of study in terms of semiotics, narrative theory, poststructuralism, gender studies, and nearly any textual methodology. Yet literary-critical studies of cybertexts have remained few in comparison to other literary forms and their now-obvious world-historical profile. Such studies as there are have sometimes been drawn off into other, newer disciplines, such as media studies, digital media, or information design. Literature, the larger sibling within English, does not offer a professional role for students and faculty to apprehend and work with.

On a more speculative note, one could consider the marginalization of science fiction within English studies. The genre remains small in terms of critical treatment, and suffers from some strong, residual critical disdain (the genre often repays the attitude, sadly). Science fiction and the Internet have been intertwined since the

ARPANET days, from the popular culture environment of programmers to one of the first great (and ongoing) virtual communities, SF-LOVERS. Science fiction texts remain rich fields of critical inquiry generally, but here can offer useful insights into cyberculture. Novels like William Gibson's *Neuromancer* (1984) and Neal Stephenson's *Snow Crash* (1992) played an influential role in the development of applications and popular Internet culture. Stories, films, and subgenres offer clues to cyberspace socialization, along with glimpses, prognostications into newer technologies, and their implications. If cyberspace appears like science fiction to many of us, at times, perhaps its generic marginalization in English limns another reason for the profession to shy away from it.

The rapid growth of social software offers an emergent threat: the loss of control over collaboration. While English instructors are skilled in creating environments for cooperative discussion and writing not necessarily using digital technology (that is, journalling, small groups, peer editing), collaborative software supports ad hoc, emergent relationships, whose nature is not necessarily bound to class space. Recent applications usually considered under the rubric of "social software" include blogs, wikis, network-building services (Friendster, Meetup, Flickr), and knowledge-sharing applications (Technorati, del.icio.us). Social software advocates urge computing as a tool for enhancing our abilities to find, share, and build trust with other people. This stems from an ideology of openness and border-crossing, unlike standard courseware restricted to an administered classroom (Blackboard or WebCT).[7] A student blogging their research in seventeenth-century literature writes for a global audience, and may receive queries and feedback from around the world, destabilizing the framework of class construction and assessment, and in turn calling into question the instructor's role as pedagogical authority or owner. The experience can be very affirming for the student, but unsettling for the English professor, whose traditional space has now been breached.

Moreover, the very recently established field of social network theory, so central to the ideology of social software, includes a controversial analysis of human interaction by scale-free distributions. These forms are profoundly asymmetrical, arguing that a minority of people (classically 20%) tends to be the network "hubs" for the majority (the remaining 80%) (Barbasi 2002; and Watts 2003). If this pattern holds in educational technology in terms of classroom discussion, campus socialization, or some other aspect of community interaction, it poses a profound challenge to our democratic sense of peer pedagogy and egalitarianism in the educational system. This sort of analysis is not restricted to social software or indeed to technological mediation, but these approaches can foreground and clarify their structures.

Again, social software returns us to a demographic divide. Our undergraduate students increasingly arrive on campus from a world including blogging, photo-sharing via Flickr, and, of course, extensive text messaging (via IM or cell phones). Social software entrepreneurs target this demographic that shape their users' information literacy. As a generation largely without future shock, they have absorbed cyberculture since puberty, and are at home in it. Which is not to say undergrads are experts, or even thoughtful about what is the informational background hum of their lives. But that lived experience is pedagogically ripe for exploration, a stratum of life ready to be activated by a canny teacher with an eye towards interdisciplinarity. Within English studies, we should be aware that our students have played stories as games many times, and with friends; such cybertextuality is an entrée to fiction, or a case study for more advanced criticism. The student as blogger is already immersed in the classic compositional tropes of audience, voice, iteration, intention, and discourse. Given advances in hardware and software, students are now likely to be familiar with multimedia both as consumers and producers. A discipline of English, steeped in multimedia study from the Book of Kells to William Blake's illuminations, is well positioned to work with such a generation.

And yet these connections are conceptual, rather than curricular, given the investment English has built up in the professor as classroom centre. Some are accessible in a small way, much as teachers have learned to reference contemporary pop culture in the classroom or in the occasional remark at a conference paper reading. To extend the profession into cyberspace is to summon a series of pressures on the identity of English studies, drawing us out from our presentation of self into something less like copyright law's imagining, and perhaps more like social software's networked identities. But in so doing we can complement the discourses of cyberspace in our research, turning the hitherto slowing friction into reflection and critique.

Notes

1. See Abbate (1999), Rheingold (2000), and Grossman (1999).
2. Curated by David Silver at http://www.com.washington.edu/rccs/.
3. First claimed as a term in "Revolutionizing Science and Engineering through Cyberinfrastructure" (2003), http://www.communitytechnology.org/nsf_ci_report/.
4. Lisa Ede and Andrea Lunsford (1990, 6), for example, note that the solitary writer image permeates "the theory and practice of teaching writing." The scholarship often depicts the writer, working alone, drawing on deeply divined personal truths or engaging in inner dialogue as the means of creating knowledge (Lowe and Williams).
5. See Vaidhyanathan (2001), Lessig (2001), and McLeod (2005).
6. For a look at one of the largest wiki explorations, see Rick et al. (2002).
7. I have not addressed copyright seriously in this paper, but should mention its role in supporting the adoption of courseware restricted to a classroom environment. The Technology, Education, and Copyright Harmonization (TEACH) Act (2000) provides legal cover for faculty claiming fair use of digital materials, an increasingly needful shield—provided the usage in question occurs in a situation limited to students in one class and bound in time to a semester or term. BlackBoard, WebCT and their ilk build in such structures at a basic level. See http://www.lib.ncsu.edu/scc/legislative/teachkit/ for information about TEACH.

3
Intellectual Property and the Humanities

Christopher Kelty

Fifty years ago, intellectual property law was an arcane and obscure branch of law governing the business dealings of professional printers. In the twenty-first century it features lawsuits against teenagers, nearly weekly articles in the *New York Times*, anticopyright art (appropriationist artists like Plunderphonics and Evolution Control Committee), and rock star lawyers' appearances on television talk shows (Lawrence Lessig). There are conferences and books, law-firms, nongovernmental organizations, technical committees, and protest organizations, plus theories enough to sate every philosophical taste and cover nearly every form of creative practice in existence. Everyone from Midwestern indie rock bands to South American traditional healers and back again seem to be discussing the finer points of copyright and trademark.

What's changed? Why is intellectual property law and practice so troubling and so important today? What role do the Internet and new media have in these changes? Why, all of a sudden, should scholars in the humanities pay attention, whether they care about new media or not?

This essay is a broad overview of these issues, intended primarily for humanities scholars who want to understand what has changed, and how, in the area of intellectual property and its intersection with the Internet and new media. It is by no means restricted to the legal issues, but takes a broader, cultural view on the subject, including the emergence of oppositional political forms within the world of intellectual property.

IP law, an overview

Intellectual property laws are generally of three kinds: patent, copyright, and trademark (a fourth kind, often referred to as *sui generis*, is commonly used to assert protection over national and cultural resources; Brown 2003). In legal thinking it is common to distinguish patent from copyright according to the so-called "idea/expression dichotomy"—that is, patents are legal protection afforded to ideas, while copyright is protection afforded only to tangible expressions of ideas. The philosophical vicissitudes of this dichotomy are rich, to say the least, and the legal literature reflects this.[1] On a practical level, however, both copyright and patent are intended primarily to grant limited monopoly rights for the commercial exploitation of "intellectual products" whether they be ideas or expressions. The issue of whether to grant legal protection to these works is debated far less than the question of which are ideas and which expressions—that is, intellectual property law is seen to be fundamentally an issue of economics ("progress in science and the useful arts" as the US constitution puts it). Trademark, on the other hand, is a body of law devoted to the use of signs and symbols that designate, distinguish and/or promote commercial products of any kind. The bulk of this article focuses on copyright, since the areas of the humanities where patent and trademark are at issue are more limited.[2] While many people find the very notion of owning a word or an idea to be absurd or even offensive, intellectual property law has traditionally aimed at something more mundane. However, the uses to which it has recently been put have become complex and increasingly frightening for scholars, artists, and citizens around the globe.

Copyright statutes can be traced back to roughly the seventeenth century and the various European efforts to control and, in some senses, censor the output of the then relatively new printing presses. The various struggles in England and the continent concerned not so much authors, but the competition of the various printing guilds and the authority of the monarchical states.[3] Often, the origin of copyright laws is marked with the passing of the Statute of Anne in 1706, which for the first time introduced the notion of a time limit on a monopoly to print a particular book. Statutory laws in continental Europe and common law in England also governed the early circulation of books between nations, but the reach of law was not absolute: no doubt many books, tracts, pamphlets, and manuscripts circulated informally throughout Europe and the world, quite ignorant of the legal constraints of diverse nations. It is important to understand that, for the last 300 years, much of this informal circulation of materials has been effectively unregulated, and it is precisely this informal, person-to-person, and noneconomic circulation that has suddenly become a space of intense regulation in the age of the Internet.

Copyright law was initially, and for most of its existence, a law meant to govern the economic relations of the sale and distribution of works. The set of issues associated with piracy, double dealing, and the propriety of printers (more than that of authors) was responsible for bringing into being the modern notion of a reliable, printed book, associated with a single author, and printed by a single publisher for a limited length of time. Such a notion, as Adrian Johns (1998) notes, may seem eminently natural today, but required a great deal of cultural and social work to be brought about. Economic intentions notwithstanding, copyright law, since its inception, has also been used as a tool for the suppression of and competition for ideas. Mark Rose tells of the first copyright case after the Statute of Anne, *Burnet* v. *Chetwood* (1720), in which the court ruled to prevent a translation of a book by Burnet called *Archaeological Philosophica* into English. The ruling did not claim that the copyright excluded translations, but that the book—an attempt to reconcile protogeology and the biblical story of Genesis—was deemed to contain "strange notions" and therefore should not be circulated in English (Rose 1993, 49–51). Through the history of copyright, the blurred line between the economic and the moral has routinely troubled authors, printers, parodists, satirists, copyists, and others.

In the nineteenth and twentieth centuries, a number of novel legal doctrines emerged that have attempted to manage issues of information ownership and circulation, such as laws concerning slander and libel, the right to privacy, and the right concerning publicity and public images of individuals, as well as a virtual "infolanche" of regulations and laws governing telecommunications, broadcasting, and information technology. Many of these laws are focused on preventing or redressing harm caused to individuals, rather than the purely economic issues of monopoly rights in reproduction—but the line is nonetheless routinely blurred. For instance, Copyright law has recently been used by the Diebold Corporation to attempt to restrict the circulation of incriminating memos written by employees (Boynton 2004, 40). The intended purpose of this action was to prevent private information, and the words and thoughts of employees, from circulating publicly—not to recover lost income from the sale of these works—and yet copyright infringement was the charge brought against the alleged offenders.

Internationally, intellectual property is subject to several conventions and treaties—most notably the Berne convention, and most recently the World Intellectual Property Organization's (WIPO) Trade Related Intellectual Property (TRIPs) agreements. The Berne Convention dates back to 1886, but has been extensively revised, with the most recent version governing copyright dated 1971. The WIPO TRIPS date to the Uruguay round of GATT (subsequently the World Trade Organization) in 1994. The TRIPs agreements are required to join the WTO, but their provisions are still hotly debated around the world, and govern everything from copyright to fair use to the trade in essential medicines. Many aspects of intellectual property law have nevertheless been "harmonized" in an effort to expand markets, and to expand the systems of protection in cultural goods as one aspect of globalization. One important difference that remains—with respect to copyright—is the notion of a "moral right of the author" present in much of European law, but absent in the common law tradition. The moral right of the author purports to govern issues that are not strictly commercial, such as the integrity of works and the requirement to identify or credit the author. Under such a doctrine, an author might therefore relinquish all commercial rights, but retain the right to claim ownership over and amend a text or work of art. In the common law case, relinquishing copyright (through sale or transfer) effectively concedes these rights as well—the author can no longer object to the uses of his or her work.

The moral right of authorship raises a related issue of relevance to the humanities, that of plagiarism. While copyright can boast an enormous amount of legal doctrine, plagiarism cannot—there is no formal law governing plagiary and a much smaller amount of legal writing on the subject, though no shortage of investigations on the part of literary scholars. Plagiarism is, however, increasingly subject to an explosion of policies at universities and colleges that attempt to set out clear rules regarding its exact definition. Many of these policies confuse the legal rights granted by copyright with the moral responsibility to give credit where credit is due—a blurring of the line that seems to give greater weight and moment to the act of plagiarism; inversely, the use of copyright law to silence or suppress parody and satire blurs the same line, and does so through the rhetorically flexible language of theft and piracy employed in both cases. Even if the intention of intellectual property law has always been the governance of the commercial realm, it has become more and more common to use it to differentiate legitimate and fair uses of works from illegitimate ones, and thus to blur the lines between the economic and aesthetic values of artistic works. In discussing the copyrightability of a circus poster in *Bleistein* v. *Donaldson Lithographic Co.*, Oliver Wendell Holmes cautioned that "It would be a dangerous undertaking for persons trained only to the law to constitute themselves final judges of the worth of pictorial illustrations, outside of the narrowest and most obvious limits . . . The taste of any public is not to be treated with contempt" (Merges 2003, 328). Yet, judges and lawyers routinely find themselves—or put themselves in the position of—adjudicating matters of taste and aesthetics, rather than law.

Important recent changes to copyright law

The twentieth century has been the century of copyright extension in the United States (as well as in many other countries, which have largely followed suit). At the beginning of the century (before 1909), protection lasted twenty-eight years, with the option to renew for another fourteen. At the end of the century protection lasted for the life of the author plus an additional seventy years (or ninety-five years total in the case of a "work for hire"). The two most important sets of changes to the US federal copyright law (US Code Section 17) recently have been the 1976 amendment to the Copyright Act and the 1998 "Sonny Bono" Copyright Term Extension Act (CTEA). In 1976, several important changes were made in addition to the term extension. The two most important were the codification of fair use rights and the removal of the requirement to register.

Fair use has been a core aspect of judicial reasoning about US copyright since the mid-nineteenth century (as well as a relatively unique one globally). Since the constitutionally mandated role of intellectual property is to "advance the progress of science and the useful arts," its scope is necessarily limited to this goal. Fair use, therefore, had long been used—informally and somewhat arbitrarily—to reason about the balance between the rights of the copyright holder, and the rights of the public or the citizenry to make "fair" use of a copyrighted work. Until 1976, only judicial precedent governed fair use; in 1976, the doctrine was codified in four fair use tests. These "tests" concern the *nature* of the use, the *purpose* of the use, the *amount* of the material used, and the *economic* impact on the rights holder. The tests are not definitive, and as a result, fair use cases have historically been just as likely to be decided in favour of the public as they have the copyright holder.[4] The politics of making fair use explicit have been very hotly debated. On the one hand, scholars and librarians have often cried out for clear rules about what is and is not fair use (for example, how much, how long, and in which contexts) and the recent TEACH Act (Section 110 of the Copyright statute) is one attempt to provide some kinds of codifications concerning use in a classroom and in distance education. On the other hand, many legal theorists and activists assert that the existing system of fair use tests provides a much greater range of potential uses, precisely because it refrains from codifying specific uses, and forces copyright holders explicitly to object to uses they consider unfair (and see them taken to court).

The removal of the requirement to register had a largely unforeseen, but insidious effect. By removing this requirement, authors, artists and filmmakers (including corporations who hired them) were no longer required to explicitly register a copyright (such as, for example, with the Library of Congress). This change meant that all written works would be automatically given the protection of law the instant pen hit paper. Though people may still informally debate about the requirements for something to be copyrighted (does it need a ©? should it be sent through the mail and postmarked?), this change to the law makes the only requirement the activity of "fixing in a tangible medium." One effect is that there is no longer any reliable or comprehensive registry of works created, as there had been, for instance, for films created prior to the 1970s. This creates great difficulties for librarians, film historians, and anyone wishing to find the author or owner of a particular work. A good example of the effects of this change is the Prelinger Archive—an archive of 48,000 "ephemeral" films (educational, advertising, scientific etc.)—some 2,000 of which are available to download. The Prelinger Archive contains films only up until 1964, the date at which it is no longer possible to determine whether a film is registered as a copyrighted work or is in the public domain.

A second and more substantial effect of this change is that it has effectively destroyed any notion of a public domain: if there is no requirement to register, and works are protected for over a hundred years, nothing can "fall into" the public domain. Though a work may effectively be "public" because its owner does not attempt to protect it (or even explicitly wishes it to be public), there is no legal way in which that status can be asserted, and a user of that work is always at risk of the author suddenly deciding to assert his or her rights. In the era of the Internet, such a fact makes a great deal of difference, as it introduces suspicion regarding every written word on every web page in existence, not simply those over which an ostensibly autonomous individual has visibly or meaningfully asserted some kind of ownership. It has changed the default condition of writing from being public property to being the property of an individual.

The 1998 Copyright Extension Act made one very important change to the law. The CTEA includes a piece of legislation known as the Digital Millennium Copyright Act (DMCA). The DMCA was intended to set legislation concerning digital works and it has been fiercely debated since becoming law, in particular because it extends new restrictions on copying digital versions of copyrighted works, and significantly strengthens the penalties of doing so (by making it a criminal felony). The DMCA's relevance to the humanities can perhaps be seen most clearly in the case of Dmitri Sklyarov and the Adobe e-Book Reader case.

In the summer of 2001, Sklyarov was a Russian programmer and Ph.D. student attending a very well-known annual hacker conference known as HeDefCon. Sklyarov was there to present a PowerPoint presentation about a piece of software called the "Adobe e-Book Reader," intended for reading a book online, or perhaps on a book-shaped device. If one "purchases" an e-book, one receives an encrypted file (very weakly encrypted, and this was Sklyarov's point) and a licence agreement that tells you what exactly you can and cannot do with this book. The terms of this contract are then effectively executed by the software, allowing a publisher to charge for different activities, such as copying the book, modifying the book, transferring the book to a different machine or device, lending the book, reading the book more than once, and even reading the book out loud. To show off these Draconian possibilities, Adobe famously demonstrated its software using *Alice in Wonderland*, a book that has been out of copyright for over thirty years. The irony was lost on no one. Sklyarov proceeded to explain how this software worked and what kind of encryption was used to prevent users from doing certain things with Lewis Carroll's masterwork. The FBI arrested him and charged him with violating the DMCA. Sklyarov was eventually released, and allowed to return to Russia, but the incident sent chills down the spines of computer scientists, security researchers, and civil disobedients everywhere (see Vaidhyanathan 2004).

The case, as well as a handful of others that have been brought as challenges to the law, raises several issues. For computer scientists, especially security researchers, it has represented an attack on academic freedom and the right to publish. For others it has raised the question of whether or not computer software code should be considered a form of speech—especially political speech—and therefore whether the DMCA constitutes a violation of the First Amendment.[5] These issues have yet to be resolved, but it is clear that one of the effects has been the widespread and sudden politicization of new media technologies and intellectual property law especially amongst a younger generation of net- and code-savvy artists and writers.

Other changes to intellectual property law seem to support the assessment that the twentieth century was the century of expansion. James Boyle (2002) speaks of a "second enclosures movement" in intellectual property. Protection (patent and copyright) has been extended to all manner of objects never before considered intellectual property, such as computer algorithms, collections of facts (databases), the "look and feel" of interfaces, genes and other biological products, business methods, and increasingly, to cultural histories and traditions ranging from myths to dance to methods and ethnobotanicals for treating disease (Brown 2003; Hayden 2003). In the twenty-first century, the intersection of the culture industries with information technology, scholarship, and science has been decidedly politicized through the apparatus of intellectual property law.

Changes in computing, networking, and writing

Changes in copyright law have long been driven by changes in media and the uses of media. The US federal statutes read like a boring legalistic version of Marshall McLuhan, filled with evidence of the rise of radio, television, cable, and satellite broadcasting—demonstrating once again that the content of any law is always another law. They contain rules governing jukeboxes, cover versions of songs, designs for boats and semiconductors; clauses that protect industry, and clauses that give libraries and educators specific carve-outs. It is little surprise, then, that computing devices and the Internet should have begun to appear in the copyright statutes as well.

Without doubt, the rise of the World Wide Web in the last ten years has been a challenge to existing intellectual property law. Consider what a change the Internet implies: in 1954, copying a book could have meant only two things—copying it out by hand or becoming a publisher by investing in expensive capital equipment to recreate the book. The former would have earned you a sore wrist, but the latter, if you were successful, may have brought you into a courtroom. In 1998, copying a book could be accomplished far, far more simply—from photocopying to digital photography to scanning, to simply copying the book from a CD-Rom to a hard drive. Indeed, computers automatically make copies of digital files simply in order to display them on a screen. The very meaning of copying has therefore changed quite radically, making it drastically easier to be hauled up on charges and accused of being a pirate, as is clear from the recent lawsuits against teenagers by the Recording Industry Association of America (RIAA).

Or, consider what it meant to distribute a film in 1954: massive investment in film processing equipment, a vertically integrated corporation with access to theatres and projection equipment around the country, shipping costs, and difficult dealings with unionized projectionists. In 2004, nearly every new computer can burn DVDs and the race to produce ever more extreme versions of copy protection technology is met at every turn with a better tool for breaking it. Projectors cost under a thousand dollars and just about anyone can operate one to create an impromptu public or private screening.

Publishing houses and film companies have reacted in various ways. Some have embraced the Internet as a marketing tool for selling books, pushing movie clips, or tantalizing consumers with free songs. Most however, under the umbrella of the RIAA and MPAA, have joined the two-pronged fight 1) to create ever more complicated copy protection technologies and 2) to lobby for legislation that makes it ever more consequential for someone to break these technologies. The most recent concept in copy-protection is "Digital Rights Management" such as the e-Book reader mentioned above. In reality, DRM has had limited success, but is nonetheless becoming more ubiquitous (Apple's iPod and iTunes service uses Digital Rights Management technology, as does just about anything built for Windows). The RIAA and the MPAA have also turned to lawsuits and to forms of propaganda, such as messages before films in theatres, and classroom materials for children that teach them not to share. (For a chilling example, including classroom materials for a Junior Achievement program called "What's the Diff?," see http://www.respectcopyrights.org).

Independent of issues of copying and distribution, one of the most significant effects of the spread of networking and computer literacy has been the blurring of the line between software and writing—both in the tactile sense that everyone now writes (at least a final version) on a screen and in a computer file, and in the less obvious sense that the distance between content (writing or creating a work) and presentation (making public or publishing a work) has been transformed. The act of "publishing" a text has become more notional than practical for anyone with access to the Web and a bit of HTML under his or her belt.

Increasingly the ease of publishing is precipitating new concerns about compatibility and archivability and about the responsibilities of software makers, libraries, and individuals to the works they create and must maintain across an ever widening array of incompatible or outdated software formats and hardware. Increasingly, students are aware of the many layers of software and applications that mediate the texts and objects they create, and are becoming more adept and creative in combining them.

Digital cameras are nearly ubiquitous, as are camcorders, and sound and video editing software. And of course, there is the perennial question of plagiarism, given new life and new surveillance possibilities by Google™, not to mention a whole new market both in term paper producing companies, and companies claiming to be able to spot plagiarists. In the end, the expansion of intellectual property law, combined with the new possibilities resulting from the Internet and the ubiquity of software and networks raises both practical and theoretical questions for the humanities.

Old and new questions in the humanities

The impact of intellectual property on the humanities can be felt in three ways: first, directly on education and writing, through issues of the fair use of texts and the IP status of humanities scholarship; second, in some of the kinds of questions that have been asked about authorship, plagiarism, and the nexus of writing, technology, and software; and third, on the disciplines and scholarship more generally, in terms of the research tools and modes of dissemination that are being challenged and transformed by the combination of new media and intellectual property law.

The direct effects on the humanities include the limitation—both legal and technical—on fair use in the classroom and in publications. The situation is obviously worse for film scholars than for literary scholars, and obviously worse for scholars of twentieth-century work than for scholars of pre-twentieth-century work. Nonetheless the issues affect everyone at some level. No scholar who teaches is immune from the question of whether or not it is legal (regardless of whether it might be legitimate) to distribute copies of a scholarly work in a classroom. For instance, scholars who wish to use film in classrooms must now face a startling fact: to create a montage of DVD clips illustrating a single scene in a Shakespeare play, a scholar must now be willing to violate the DMCA multiple times—an action that could potentially bring a fine of up to $500,000 and five years in jail. The alternative, as many university counsels have begun to advise, is to begin an arduous process of getting permission for each clip. And while such advice might address the legal issue, it underestimates the greater technical difficulty of obtaining from entertainment companies usable footage in a format that can be manipulated by a reasonably proficient computer user. Fortunately, some responses and alternatives have begun to emerge, and they are reviewed in the last section of this essay.

In terms of changing scholarship, the issues are much more interesting, and arguably less threatening. The question of authorship has occupied literary and cultural studies with renewed vigour since the publication of "What is an Author?" by Michel Foucault. While the debate concerning this work was not initially about intellectual property law, several scholars have taken up the question of what difference it has made. A symposium and edited volume (Woodmansee and Jaszi 1994) introduced a raft of issues related to the rise of copyright and the notion of originality and creativity in European authorship. Mark Rose has written a detailed work on the earliest cases of copyright litigation (1993). Siva Vaidhyanathan (2001) has traced the cultural and literary intersections of the public sphere and copyright law, and Paul K. Saint-Amour (2003) has contributed an excellent volume on the subject of economic and aesthetic value from the early nineteenth century up to the work of Joyce.

The authorship question has similarly invoked questions about forgery, plagiarism, and piracy. Christopher Ricks (2002, 219–40), for instance, has nicely summarized one aspect of the debate about plagiarism and copyright: on the one hand, there are proponents of an historicist view of plagiarism and authorship (that it might have emerged with the growth of laws and practices of copyright), and on the other, proponents of a more morally absolute position (perhaps closer to the "moral rights" legal idea) in which the definition of plagiary has never historically been in question.[6] Both positions are well represented in the scholarship, but the distinction between legal (formal) and moral (informal) spheres of creativity has only recently become a core question of this scholarship. In *The Nature of the Book*, Adrian Johns (1998) pays much closer attention to the details of both the legal transformations and the actual workings of "print culture" in seventeenth- and eighteenth-century Europe. Johns' work is a corrective of sorts to the strong form of arguments about print culture made by Eisenstein and McLuhan.

Parallel to scholarship on authorship has been a vibrant field of investigation into the blurring lines between writing and digital media. Many early works explored the concept of hypertext and the limits and the meaning of electronic writing (Landow 1992; Bolter 1991; Joyce 1995; Heim 1987). Here, traditional authorship may be questioned, and the proposed radical nonlinearity of hypertext may well produce linkages and hitherto unknown conceptual possibilities—but copyright law has largely gone uninterrogated in this area. Even recent works that propose to draw together classic approaches in literary, philosophical, and cultural studies with close attention to new forms of media and their constitution (for example, Bolter and Grusin 1999) are almost uniformly silent on the subject of intellectual property.

Part of this reluctance is no doubt the perception that intellectual property is an economic and legal phenomenon not a cultural or social one. Though the case has been made that the proper place to look for the meaning of an "information society" is law (Boyle 1996), most scholarship is hesitant to grant to law that kind of power. Recently, however, a number of events and emergent areas of art and scholarship have challenged these assumptions. Chief among these have been increasingly high profile court cases that have engaged an increasingly large segment of Internet activists, geeks, artists, scholars, and writers such as the eToy case in 1999, when a Swiss art group eToy was sued by toy retailer e-Toys over trademarks and an Internet domain name (Wishart and Boschler 2002). Similarly, cases involving copyright infringement, academic freedom, and free speech have given scholars reason to reconsider the relevance of intellectual property law to the transformation of culture.

The 1990s saw the rise of a new generation of critics, artists, and theorists who focused their attention on the political and cultural effects of the Internet and software (for an introduction, see Lovink 2002). The most creative blurriness has occurred at the interface of software code, art, and poetry. The *transmediale* art festival in Berlin has given a prize for best software since 2001, and there have been several projects that investigate the boundaries of software, copyright law, and art (see Weibel and Druckery 2001). The ambivalent implication of much of this work is that intellectual property is a necessary—but not intransigent—evil to be dealt with creatively and in a manner that does not reduce solely to the resigned hiring of lawyers.

Finally, the new research tools available in the humanities have also been forced to confront intellectual property issues head on. The availability of material in repositories such as JSTOR (http://www.jstor.org/), The Valley of the Shadow Project (http://valley.vcdh.virginia.edu/) and an increasing number of full-text and image archives has meant both that work is more available than ever before (and scholars more beholden to consulting it, perhaps) but also more legally and technically restricted than ever before. Such projects routinely struggle with both the technical requirements of making work available, and the demands by university administration that they produce revenue in addition to scholarship. This seemingly paradoxical situation has increasingly led scientists and scholars to call for a reassertion of the free circulation of scholarship, and to seek out alternatives that allow them to freely circulate information without giving up all control over it.

Alternatives to the current intellectual property system

While the current legal environment of intellectual property may seem vibrant—or dreary—enough, it is not where the bulk of the action is. A number of alternatives to the current intellectual property system—ranging from radical to pragmatic—have emerged in the last twenty years, and gathered intense momentum in the last five. There are three worth mentioning here: the free software and open source movements, the Creative Commons project, and the open access (or Free Online Scholarship) movement.

The free and open source software movements (FOSS, or occasionally internationalized as FLOSS—Free, Libre, and Open Source Software) have been the most influential alternative IP movements in recent history. Free software ("Free as in speech, not as in beer") is copyrighted software that comes with a licence that allows the user to copy, modify, and use the software however he or she sees fit. It also gives the user the right to redistribute the software so long as it is distributed along with the original licence (which gave the user this right in the first place). Hence, it is "reciprocal" (opponents say "viral") software. The implication of this is that it creates what lawyers like to call a "privately ordered legal system"—or a system in which federal or state law is ignored in favour of mutual agreements to abide by private contracts. Of course, free software depends on federal law for its definition and on a judicial system for its enforcement in extreme cases, but it nonetheless has managed to exist successfully and relatively peacefully for about twenty years. The first such "free software licence" and still the most widely is the "General Public Licence," created by Richard Stallman (uebergeek and founder of the Free Software Foundation in 1984).

Free software is a "movement" for a couple of reasons. First, it has successfully produced some very high quality software (such as the GNU/Linux Operating System and the Apache Web-server) created and used by a very large number of people, both individuals and organizations, few of whom are formally employed to do so. Second, it is built around varying commitments to the idea of freedom of information: from radical freedom ("all software should be free") in the case of the Free Software Foundation, to more pragmatic, business-oriented freedom (which is the role of the Open Source Initiative, www.opensource.org, who see themselves—not always coherently—as selling the idea of free software to capitalists). Most free software users tend to be proselytes, encouraging others to make use of free software. Several large corporations (notably IBM and Apple) have also "gotten religion" and promote free software both because of its moral advantage, but more importantly, because it represents a significantly cheaper model of software development than has hitherto been practiced.

The success of the free software and open source software movements has in turn caused people to consider the implications for other kinds of digital content: music, movies, software art, text, or photography. Several attempts to create free software-like licences for nonsoftware culminated in the launch of the nonprofit Creative Commons (CC). CC is an organization founded by Stanford law professor Lawrence Lessig, modelled explicitly on the Free Software Foundation. CC provides a set of licences that people can use when they create works that they want to allow people to use, perform, display, or distribute, or even to modify or extend. The explicit goal is to (re)-create a public domain—often referred to as a "commons" since the work remains copyrighted, but limited uses are allowed—and the tag line is "Some Rights Reserved." CC is a more of a pragmatist organization than the Free Software Foundation—but neither is explicitly opposed to intellectual property law itself. Both see problems only with the current configuration of the laws and seek ways to "opt out" of that system in favour of one that gives individuals the rights to which they are willing to agree, rather than the ones that entertainment corporations wish they would accept.

Creative Commons has encouraged several kinds of experiments. Founder Larry Lessig released his third book *Free Culture* under a CC licence on the Internet, and encouraged readers to transform it in limited ways: within days there were audio versions of the book, translations, and reconfigured web-based versions. Similarly Cory Doctorow, a science fiction writer and digital rights activist, released his novel *Down and Out in the Magic Kingdom* both as a print book and as a freely downloadable file. A year later, he re-released the same book under the most generous licence terms (allowing modification and commercialization) and his work has undergone a similarly enthusiastic set of transformations (http://craphound. com/ down/).

Community-based projects based on Creative Commons licences have also emerged, such as Opsound (an artist-organized community of people who share and reuse sound samples), Openphoto.net (which allows people to post and download photos), and fourthwall (a project to release all of the raw materials for a film as an experiment in "open source cinema" http://fourthwall. creativecommons. org/).

With the availability of these "open content" licences, universities have also started to consider their use in innovative ways. One clear example was the announcement by MIT that it would release all of its teaching and course material online as part of an "open courseware project" (http://ocw.mit.edu/). A similar but more radical project is Rice University's Connexions Project (http://cnx.rice.edu/), which allows scholars to publish their educational and teaching materials in small chunks (modules) under CC licences and to recombine contributed materials into new courses and new textbooks for use in classrooms.

A third related alternative to the current intellectual property regime is the so-called "open access" or "free online scholarship" movement (an excellent resource is Peter Suber's exhaustive web site, newsletter, and blog at http://www.earlham.edu/~peters/fos/). The open access movement seeks to ensure the continued and easy availability of scholarly publications either through "self archiving" (in which scholars deposit their works in online institutional repositories or on their own web sites) or through open access journals that do not restrict access to published work. The movement has steadily gained momentum despite the warning cries of the powerful scholarly publishing corporations (such as Elsevier and Springer) that it is economically unsustainable and will lead to the proliferation of worthless research online. Despite these claims, a number of open access journals and repositories have appeared (both public and private) and some even appear poised to become self-sustaining, if not profitable (see http://www.doaj.org/ for a list). In addition, federal governments have begun to take note. In August of 2004, both the US and UK governments considered proposals to require open access to all federally funded research. In the US, the first such requirements are likely to be applied to research funded by the National Institutes of Health, and made available in their open access journal *PubMed Central* (http://www.pubmedcentral.nih.gov/).

Conclusion

There are a lot of open questions in the world of intellectual property. Some are pragmatic and political: can national legislation be made to balance the interests of the entertainment industry and scholarly and artistic cultures? Are the alternative projects that make use of private contracts and free circulation sustainable—legally or economically? And what kinds of specific legislative changes are needed? Many of these issues are addressed directly by groups like Public Knowledge (http://www.publicknowledge.org) and the Electronic Frontier Foundation (http://www.eff.org). Other issues at the local and national level may have few supporters to defend them from the powerful and experienced lobbyists for film, music, and publishing industries.

At another level these changes seem to suggest yet another set of changes in the meaning and concept of authorship. Increasingly large collaborative projects (films, documentary projects, encyclopedias, and so on) are facing the question of how to distribute both the work itself, and more importantly, the *credit* for work. Similarly, works are increasingly open-ended and reusable (sampling, mash-up, extensible software and social art projects and remixes, collaborative encyclopedias, and new forms of scholarly collaboration across cultures and disciplines)—how will credit and value be conceived in these works? What kinds of conflict might emerge between artists committed to versions of interior genius and those interested in leveraging enormous, anonymous or self-replicating forms of artwork?

At a more pragmatic level, there are many things humanities scholars should be thinking about. In the classroom, scholars should challenge and resist the ever-increasing demands to ask permission for uses and transformations that are a crucial and routine part of teaching students. Scholars should assert and exercise their fair use rights (where they exist) in all cases they judge to be appropriate, seek permission in those cases where it is easy (or economically feasible) to do so, and scrupulously give credit where credit is due. In research, scholars should be aware of the implications of the contracts they sign—and demand from publishers that they allow authors either to maintain their copyright in the work they produce, or maintain the right to allow open access to the work, especially in cases where federal funding was given. Members of professional societies (especially those who publish scholarly work) should bring these issues to yearly meetings and governing boards and demand a solution that does not sacrifice the circulation of scholarly work in favour of revenue. And for those scholars who make their work directly accessible via the Web, because they wish people to read, cite, or teach with it, they should say so explicitly (perhaps using an open licence like those provided by Creative Commons).

In the end, if there is something transformative of society in the rise of new media, it comes not only from the changing experience of reading and writing but the changing conditions of circulation, technology, and law as well.

Notes

1. One of the standard US textbooks on IP law is Merges, Menell, and Lemley (2003), which directly addresses the intersection of technology and IP Law.
2. Patents can interfere with particular software technologies that may be relevant to work in humanities computing, and trademark has been widely analysed in the realm of commodity aesthetics and cultural studies (see especially Coombe 1998).
3. A classic of US copyright history is Patterson (1968). On UK history see Feather (1994).
4. See Lawrence and Tinberg (1989) and for a more enjoyable tour Negativland (1995). Georgia Harper and the University of Texas have created a "Crash Course" in intellectual property that covers many of these issues in more detail: http://www.utsystem.edu/ogc/intellectualproperty/copypol2.htm.
5. See, for instance, the Electronic Frontier Foundation's archive of materials on this issue (http://www.eff.org/IP/DMCA/), especially the Edward Felton Case, and the 2600 Magazine Case, both of which were dismissed before they could challenge the law.
6. Ricks (2002, 226) uses as his evidence of the timelessness of plagiarism the etymology of plagiary, from *plagiarus* (L. kidnapper), and that it is used in several poems by Martial in which he discusses the theft of his work. On this debate over plagiarism before copyright, see also Kewes (2003).

Part 2

A New English? Theory and Practice

4

Putting IT into the English Syllabus: a Case of Square Pegs and Round Holes?

Stuart Lee

Anyone reading this book may already be defined as "one of the converted"; that is, someone who believes that there is scope to introduce IT into the teaching and study of English. They are perhaps looking to the chapters presented here as a guide to how this could be successfully achieved. Yet, such beliefs may still be seen as representing a minority view—hence the title of this contribution—forcing a new medium into a discipline without any just cause or reason. This essay therefore presents a discussion centred on the introduction of IT into the English Literature *syllabus*, not as a means to assist in the teaching of literature, but more as a subject in its own right. This chapter contends that many of the emerging areas we are witnessing within IT are natural bedfellows to the discipline of English studies. Moreover, English is in a prime position to seize these areas as its own; and, more importantly, if we fail to do so the consequences could be serious in terms of our credibility with our students and with colleagues in other disciplines.

English studies and IT: a natural synergy

Let us begin by considering the key skills we expect of or attempt to instil in our students. Good guidelines would be those identified by the UK's Quality Assurance Agency in its *English Benchmarking Statement* (2000). Here the core skills for the discipline are defined as:

- *critical skills in the close reading and analysis of texts;*
- *ability to articulate knowledge and understanding of texts, concepts, and theories* relating to English studies;

- *sensitivity to generic conventions and to the shaping effects upon communication of circumstances, authorship, textual production, and intended audience;*
- responsiveness to the central role of language in the creation of meaning and a sensitivity to the affective power of language;
- *rhetorical skills of effective communication and argument, both oral and written;*
- command of a broad range of vocabulary and an appropriate critical terminology;
- *bibliographic skills appropriate to the discipline, including accurate citation of sources and consistent use of conventions in the presentation of scholarly work;*
- awareness of how different social and cultural contexts affect the nature of language and meaning;
- understanding of how cultural norms and assumptions influence questions of judgement;
- comprehension of the complex nature of literary languages, and an awareness of the relevant research by which they may be better understood (emphasis added).

Along with the generic skills also presented in the guidelines, the above represent an attempt to encapsulate the learning objectives and particularly the outcomes one would expect to find from any degree in English within the UK; and one suspects that many of these recommendations would have international applicability.

Let us examine these skills more closely, and in particular those that have been emphasized for the purposes of this discussion. Although the authors of these guidelines clearly had in mind the study of literature via the printed medium, it is suggested that these could apply equally well to text in any form (including, for example, electronic text). Moreover, in combination they more or less present the skills needed to analyse and criticize the production and dissemination of electronic texts especially as they appear on the World Wide Web. The Web after all is predominantly text authored by literate people wishing to communicate some information, or to provide access to a set of resources. Regardless of whether the web site is commercial or not, all the interfaces and navigational options are presented in a way that allows readers to interact with the information in a certain way, to explore or to follow directed paths, and to engage with the material—all packaged in an intertextual world of hyperlinks. This description is perhaps not too far from that of literature itself. It is not surprising therefore that two well-known theorists behind the "web"—Ted Nelson and George Landow—have literary backgrounds.

Taking this observation further, the Web is quickly becoming one of the prime sources of information for most students, and increasingly for many researchers (notably as more and more primary texts become available to supplement the existing bibliographic tools). If we put these two concepts and developments together then it is somewhat peculiar that English departments around the country are not seizing the Web and the study of the Web as their own. To quote a recent English Subject Centre report:

> The possibilities for the application of IT in English are potentially extensive since the advent of the World Wide Web (WWW) has resulted in a proliferation of textuality. English is pre-eminently placed to engage critically the textual and discursive production that attends the Internet Age. The subject as a whole, however, has not embraced this opportunity. (Hanrahan 2002, 3)

English studies and IT: the requirements

When considering why English departments might wish to consider adopting the study of electronic texts and the Web into their syllabus, one may persuasively argue that they have very little choice in the matter. There are three areas that could be seen to be enforcing the adoption of IT into the English syllabus. The first relates to the acceptance of what it means to be "literate" in the twenty-first century. Undoubtedly the use of IT is now being recognized as a basic literacy requirement for all professional, and many social, activities. It has not quite reached the status of numeracy or standard literacy, but that time is fast approaching. If we explore further what the definition of literacy is as related to IT then again we can see parallels with the English discipline. Elizabeth Daley, Dean of the School of Cinema-Television at the University of Southern California, has proposed a good model for this. In her article "Expanding the Concept of Literacy" (2003, 33–4) she suggested that:

1. The multimedia language of the screen has become the new vernacular.

2. The multimedia language of the screen is capable of constructing complex meanings independent of text.
3. The multimedia language of the screen enables modes of thought, ways of communication and conducting research, and methods of publication and teaching that are essentially different from those of text.
4. Lastly, following from the previous three arguments, those who are truly literate in the twenty-first century will be those who learn to both read and write the multimedia language of the screen.

Daley's definition of literacy is attainable, and indeed many students are already there. Furthermore if we accept her definitions then we also support the idea that all our graduates must achieve these in some way before they progress into the job sector or postgraduate study. Yet is it the role of the English department to instil or check these new literacy skills? Perhaps not. Perhaps we should simply rely on preUniversity education, or support services at our institutions. However, rather than looking on this matter negatively, we should consider it as an opportunity to bring a new area into the fold of the English syllabus in a manner that is appropriate to the discipline.

However, the argument as to why we have to bring IT into our teaching does not stop there. Consider, for example, that all the jobs that our students will be going into after graduating will require IT skills. Yet in a report by Brennan and Williams (2003), the authors noted that one of the key areas in which English graduates seeking employment felt they were at a disadvantage compared with other graduates was their IT skills. What makes this even more alarming is that we often say to our prospective students that they will be trained in these key areas. Brennan and Williams (2003, 24) surveyed several English Department web sites and found that many of them, at some point, stated that students would acquire IT skills during their degree. Of course, this may happen through enforced training using the institution's IT services but there is very little evidence to show that this is taking place (especially in the UK). More importantly though, are we making any efforts to make this relevant to literature students? To quote a recent report on *IT and English*, prepared by the English Subject Centre:

> While most students of English know how to word process, use e-mail and probably make use of limited search facilities, the integration of the new technology into the experience of English

> education nationwide is restricted. The HEFCE summary report of the Quality Assessment round in England (93–94) . . . highlighted this area as a deficit in English. (Hanrahan 2002, 3)

There are three solutions to this problem. The first is simply to make more use of electronic resources in the teaching of literature and, more importantly, to train students on how to use these resources properly. If we consider the wealth of subscription-based resources in the areas of American and English literature, plus the free web-based resources catalogued by such excellent services as the HUMBUL Humanities Hub or Alan Liu's *Voice of the Shuttle* then this is not too onerous a task. In doing this, students will become familiar with the Internet and accept it as another resource at their disposal, which is as familiar to them as the library. There are also online tutorials specifically available for literature students (such as the Internet for English site), which train them quickly and easily in finding and assessing literary resources. These resources can now easily be embedded as course support material via Virtual Learning Environments such as Blackboard, WebCT, or Nathan Bodington.

The second is to create resources (such as a web site) to support a particular subject. If we accept that the tutorial on Isaac Rosenberg mounted in January 1995 was the first web-based teaching package for English, then there is over ten years of experience and examples available to inspire and inform (see the "Virtual Seminars for Teaching English Literature" Project). The main advantage of doing this is the ability to produce something that specifically targets local teaching needs, but this comes at a price, and that is the time needed to create the resource, which can be extensive.

The third option, and the one which will be explored in this article, is to consider an addition to the syllabus—namely modules aimed specifically at the interaction between English and IT, in a manner that is both applicable and appropriate to literature students. What follows then is an example of one such course, the E-Lit option currently offered at Oxford University.

Introduction to the course

At the University of Oxford third-year English literature students are offered a series of options, one of which is E-Lit: IT and English Language and Literature, an assessed course with the marks ultimately going towards their final honours. Understandably, the course is relatively new, it is now only in its third year, but it illustrates some of the points made earlier. In short, it aims to introduce students to the potential (and realized) uses of IT in the study, publication, and authorship of English literary works and English language use. In other words this is a direct example of the type of course that could bring together the IT needs of the students and employers with the discipline of English, and at the same time keep to the key skills outlined in the English benchmarking statement.

The structure of the course is set at six 1.5 hour classes, followed by three weeks unsupervised work on the student's project (which is to be submitted in the ninth week of term). The learning outcomes for the course are defined as attaining an understanding of:

1. How new technologies directly impact the way people research and teach English literature and language;
2. How the world of writing and publishing is changing because of the information age, plus obtaining an overall view of the current electronic publishing industry;
3. The history of the main computer-based English literature/language projects from the earliest ideas in the 1940s up to the present day;
4. What English literature/language resources are available on the Web and how to evaluate them;
5. What a hypertext novel is, and what an electronic book is, and how they are written, read, and published;
6. How the computer can help in analysing literary texts (that is, authorship studies), and equally important how it can be used to help in the study of the English language;
7. Essential transferable skills that will be invaluable when seeking a job or research position in academia, the media, publishing, teaching, and so on.

Specifically they had to able to: use most of the computer applications employed by literary scholars; use the Web effectively; create a web site (and an example to show prospective employers); design web pages (at an introductory level); and evaluate web sites.

The course is assessed in two forms. During the six weeks of classes the students are asked to submit two essays, these are commented upon, but not graded as such. The first one usually takes the form of selecting a web site or an electronic resource related to English literature, and the students are asked to critically evaluate it using guides presented as well as their own skills acquired during the previous two years of study. The second essay concentrates on the use of electronic text analysis tools to pursue literary and linguistic study. The course is generally taught by three lecturers and, outside of class, contact is maintained via a bulletin board within a virtual learning environment (AKA Learning Management System).

The second form of assessment is summative, and is the part that is actually graded. On the sixth week of the course the students are presented with a list of "topics." They are requested to choose a topic and to build a web site around it aimed at a first-year undergraduate audience. In addition they must supply a 3,000 word essay/report to support the site. The set text for the course was Condron et al. (2000). Although this is somewhat dated now, the short essays contained in the book do still offer a valuable introduction to the subject.

The existing course: perceived good and bad points

The course design had to achieve learning outcomes that matched the aims noted above, and led to assessments that evaluated these aims. On reflection this design seemed to work well (though some modifications were initiated as a result of student feedback). The reason for its success is due to the fact that the course was designed as a collaborative effort among the three lecturers based initially upon a sequencing of the content and listed under six possible topics. Once completed, we designed the final summative assessment. It was then a matter of working back through the structure to see if the classes led effectively towards the assignment.

The classes themselves worked well on the whole as all three members of staff introduced as much student activity as possible, avoiding passive engagement with the information. The opportunity to provide a hands-on element using the new technologies greatly facilitated this (the room itself had a suite of computers we could use at any time). All classes began with an opening discussion, a short talk, hands-on sessions (often in groups), and then class discussion led by the students.

Another important point of the classes was that they all kept to a "learning spiral" (Northedge and Lane 1997, 21–2). The opening discussion in each class introduced the key concepts. The students then performed hands-on work. The class reassembled and then discussed the ideas. Because the groups of students were asked to look at different products, the observations they came up with were often unique (but equally valid) which began the spiral again. The groups were also encouraged (outside of class) to look at more products to fine-tune their critical skills and discuss these on a web-based bulletin board.

Perhaps the main strength of the course (which came through clearly in the feedback) was the amount of effort put into student support. Prior to the course an audit of student skills was taken. This established who the students were—including their backgrounds and prior knowledge—and highlighted any problems that may arise. If an issue arose it was tackled directly: for example, students with insufficient IT skills were directed to a free introductory web-authoring course. The interview also presented a good opportunity to talk to the students. Most of them had thought of a career in publishing and thus a visit to a publisher was arranged as an extracurricular activity.

A common criticism from students on other courses is that it is often hard to find out all the information relating to the classes. For this reason a course web site was created within Oxford's Virtual Learning Environment known as "WebLearn" (powered by the open source system Bodington developed at the University of Leeds). This site presented a reading list (required reading and recommended reading, with associated shelfmarks) that was updated weekly with the lecturer's slides to minimize note-taking in the class. Moreover, the above mentioned web-based bulletin board was not only used to promote discussion but also to answer student questions. This helped develop a "learning community."

There were also several negative observations of the course. First, the flexibility in the overall design was greatly constrained by what is "normally taught" in similar papers: that is, fitting the context of what was expected of the third-year syndicated options in terms of class time, assignments, and weekly activities. When it came to the summative assessment we also encountered problems. The method chosen was that the students were required to create a web site and present an accompanying report. This raised serious issues for both assessors and students. As Laurillard (2002, 206) notes: "The kind of work students do using learning technologies is necessarily different from what they do in learning via other methods. Therefore the teacher has to decide on what counts as a good performance, and what counts as useful feedback." Although we felt that the learning cycle of the classes had led to deep learning of the concepts (following D. A. Kolb's discussion of experiential learning [1984]), we perceived anxieties amongst the students as to the criteria by which their web site would be assessed and the weighting of the marks. Our weekly reflective diaries all record questions from students along these lines, and the bulletin board contains several messages from students seeking further explanation. It is worth noting at this point that this took us somewhat by surprise, even though it is clear from the research into this area that students regularly focus directly on assessment criteria.

To tackle this we produced a set of evaluation criteria for the students. For the web site, we identified the following areas: the structure of a web site adheres to design principles and demonstrates knowledge of accessibility issues; it reflects knowledge of other relevant sites; its content is accurate and its tone reflects target audience; it is functional and reliable (for example, the pages are available and all links work); and it makes innovative use of emerging technology. For the report, we established the following requirements: that it contain a clear statement of topic and target audience; a discussion of the process by which the site was constructed in the light of evaluation criteria used for assessing web sites; a review of relevant literature; an awareness of contextual issues (technical requirements for clients to view the site, accessibility issues, copyright problems); suggestions for future development; clear presentation; and a full, accurate bibliography, including links to other online resources. These guidelines still left a question unanswered for the students—"how big should our web site be?" With an essay, word length is the standard limiting factor that students are used to, but here the figure of, say, "3,000 words" was meaningless. In short our answer reflected the nebulous nature of web sites and we could only give rough estimates of guidance: namely, fifteen pages for the site.

Evaluation of the course

As this was a new course it was deemed essential that a thorough evaluation of it was undertaken to see if some of the claims made earlier in this article (and assumed at the launch of the course) were valid. A variety of methods were employed to assist in the evaluation: online student feedback questionnaires; peer observation by colleagues who were asked to sit in the class and observe any problems or issues and suggest possible improvements; and internal review at the end of the course, drawing together personal teaching logs, and meeting with colleagues to discuss experiences. These evaluation methods yielded a considerable amount of information, which the three lecturers teaching the course analysed. In Table 4.1 below, the major points emerging from the above evaluations are noted in the left-hand column, and our interpretation of the replies are noted in the right.

The above interpretations led to some conclusions that have had direct effects on the structure of the course. It is perhaps easiest at this point to concentrate on the main issues that came out of the evaluation, and to indicate what alterations have been made to combat them.

The success of the classes was probably due to the notable presence of student activities in each lesson. These breaks within the lectures also allowed engagement between lecturers and students and the opportunity for collaboration between students. Bligh's observations

Table 4.1 Interpreting course evaluations

Replies	***Interpretation***
Course academic level marked as "about right" or "average."	On the whole the course was matching the level expected of a third-year option.
Student support, lecturer involvement and enthusiasm marked high.	Successful use of web site to support the courses, and the bulletin board. Topics covered matched lecture's own expertise and areas of interest. Students react favourable to al lecturer that shows "interest" in the subject they are teaching.
Timing and coverage of final assignmen was problematic.	Students had difficulty in creating and designing a web site from scratch in three weeks. Students were unaware that all topics covered in classes could appear in the summative examination.
Module on "e-books" was not well received.	This topic, which on paper looked like a larger issue, in reality did not justify its own class.

(1998, 14) that "discussion methods" promote thought and that "at some stage the student should practice it" are especially apt for the E-Lit option. Moreover the classes used mixed methods (lecture, hands-on practical analysis, and discussions), which sought to accommodate the different types of learning patterns one might observe in the students.

As noted above, students found difficulties in terms of the final assessment. Some of this is impossible to alter because the course regulations are quite strict as to how and when final assignments should be submitted. Nevertheless, one immediate change we made after the first year was to be more explicit about the criteria by which the summative assessment would be marked. We also decided that more time should be made available in class to prepare the students for the final project, and due to restructuring of the course the final lesson was set aside to do this in depth.

Summary

At the beginning of this essay I made some bold observations about the need for English syllabi to adopt IT as part of mainstream teaching. The E-Lit option outlined above and currently being offered at Oxford is a real-life example of how this might be done, and this example can now be backed up with student evaluations (see Appendix). Yet it is clear that not all faculties and departments could accommodate such a course, or have the expertise or facilities to teach it. Therefore, to many the above option may be unachievable. Elements of the course could nevertheless appear as part of more traditional subject areas. With the proliferation of computers now in universities, and the ease by which students can access the Internet, it would be a relatively easy step to occasionally ask students to review online products as opposed to basing their work solely on printed resources. The key to this is in part imagination by the lecturer, but also a long overdue recognition that digital resources have as much validity as printed resources, and should be viewed as being on an equal footing. There may not be as many digital resources of quality around at present relative to printed books and journals, but this is fast changing.

Appendix: student feedback

The feedback from the course bears out the objectives and learning outcomes outlined at the beginning of this essay. Some samples are:

"It was exciting using my critical skills to analyse a web site."

"I was amazed at what I had managed to achieve in just three weeks. For me, this was the best bit of the course, because it allowed me to combine research with creativity. It is satisfying to have completed a project and be able to share it with others and show it to future employers."

"This was an enjoyable and modern course, which teaches some important transferable skills. This is the way Oxford English options should be going in the future. I think it produces more well rounded and well informed English graduates."

"A large part of my reason behind taking the course was to really challenge myself to make use of and interact with a technology which I find quite alarming (!), so it was a challenge, and an unfamiliar working environment, but an exciting and profitable one. A number of times during the course I thought if only I'd known about these resources two years ago."

"There could be more integration of the use of web resources earlier on in the course—perhaps encouraging more lecturers to make use of web sites in notes and references."

"The course is really well taught and because it is relatively new it really helps to have the support that the bulletin board offers and having tutors willing to answer questions etc. all the time is great. It's quite a major difference from the normal system where you generally don't have much contact with a tutor from one week to the next. It's also good to keep the discussion going throughout the week with other people doing the course."

"I really enjoyed doing the course and it will certainly come in handy in the future, in a way a number of my papers probably won't."

5

A Technology of Our Own: the Place of Computers and the Case of the Small Press

Jim O'Loughlin

I

Borg or cyborg? Too often, discussions of the role of computer technology in English studies boil down to one of these unfortunate metaphors. From one perspective, the study of English risks losing that which makes it distinctive and vital if it adopts technologically-focused methods of instruction. The fear in this case is that technology acquisition by students could become the measure by which English departments are judged, and English departments would fare about as well as an undefended planet "assimilated" by the Borg in *Star Trek*. On the other side of this debate, computer technology is posited to have utopian and revolutionary potential. The possibilities for creative and critical endeavours opened up by the World Wide Web and assorted new computer programs seem endless. From this perspective, the prioritization of print and print culture in English departments risks rendering these departments irrelevant and hampers the work of those who embrace the position of "cyborg," making new technologies fundamental to the work they do. For people on the digital side of the divide, the "tyranny of the page" must be overthrown.

As opposed as these two approaches to technology first appear, what they both have in common is an understanding of computers as some kind of alien force, either malevolent or benevolent, that confronts English studies with a wholly unprecedented situation. In this essay, I argue that the issue of computer technology in English studies, while historically unique, is not without precedent. The technology of publishing has long played a vital role in literary history.

However, because literary study has traditionally focused little attention on the technology of publishing, current developments in computer technology appear as some sort of paradigmatic break. A notable exception here is recent work in Book History that examines print culture as a historically specific occurrence. Intellectually, there are many shared concerns between those working in New Media and those working in Book History, but to date there has been little convergence between the two fields. While literature is by no means the main reason to use technology, it is one reason to do so. And by emphasizing those reasons—by placing the technology of publishing within the mainstream of literary history—technological developments become part of our inquiries into any literary endeavours.

In examining the position and possibilities for computer technology in English studies, I will turn to recent work by Jerome McGann and William Paulson that avoids both the Borg and cyborg formulations while attempting to think through the fundamental impact of computer technology on work in the humanities. Finally, I will discuss in detail my experience teaching a course, The History and Practice of the Small Press, that offers a practical example of some of the theoretical concerns of this essay.

II

It is difficult to think about the place of computers in English studies apart from what has been termed "the crisis of the humanities." As the beginnings of this crisis preceded the wholesale adoption of computers on university campuses, computers cannot be held accountable for the crisis, though debate continues as to whether computerization has exacerbated the crisis or offers a potential solution to it. Regardless of the role played by computerization, most commentators agree that over the past few decades the humanities has faced declines in stature, in numbers of majors, and in overall funding.

The dilemma for literary studies has been laid out in stark terms. As Bill Readings (1996, 85) put it, "the field of literature as such is currently structured in institutional terms that are neither practical nor ethically defensible." For John Guillory (1993, 45–6), the problem has been that the study of literature no longer provides the kind of "cultural capital" that is deemed necessary for the professional-managerial class. The cache of computer-based knowledge, however, remains high, and poses both possibilities and problems for those who use them in literary studies.

When I first began offering literature courses with a computer instruction component, I thought of this work in defensive terms. At the time, I taught at a college where Engineering and Business were the dominant fields, and the English department was constantly being measured according to standards set in dramatically different areas. In such a situation, the argument that literature courses could attract both students and general interest by making better use of technology was a persuasive one. As students and much of the administration increasingly viewed computer literacy as essential for postgraduation success, traditional literature courses faced a problem. If computer instruction was wholly separate from a literature curriculum, the implicit message would be that the work of literary study was *not* part of the knowledge students (rightly or wrongly) felt they needed to get on in the world. But if students were interested solely in technology instruction, they would not sign up for English courses in the first place. The trick was to develop courses that did not simply tack on isolated training in software programs (programs which would most likely be obsolete by the time students graduated) but which provided a discipline-specific context within which to develop and continue to use meaningful computer skills.

The final section of this article details a course I developed under these circumstances. And while I stand by the scope and purpose of that course, my arguments for it today would be much less defensive. Drawing upon recent critical work, I have come to be persuaded that computer-based courses should be thought of less as a supplement to a literature curriculum than as an emerging focus likely to influence English studies as a whole. Importantly, I now believe that this innovation will be driven, not by outside pressures (as I earlier felt them), but by developments internal to the fields of literary criticism and pedagogical research. Computers stand to provide "a technology of our own" to the extent that we recognize their creative uses as a part of, as opposed to apart from, literary history.

One work that has been vital to my rethinking of this issue is Jerome McGann's *Radiant Textuality* (2001). In this study, McGann draws upon his experience as editor of the hypermedia research archive, *The Complete Writings and Pictures of Dante Gabriel Rossetti* (http://jefferson.village.virginia.edu/rossetti/), to argue for a junction of the literary and technical. In his work with the archive, McGann, despite some initially hostile responses, has found computer technology to be essential to working on the kind of issues raised by contemporary literary and cultural criticism. Rather than seeing computers as some kind of outside alien entering the field of English studies, he views their introduction as part and parcel of other innovations of current thought.

> The significance of the changes being wrought through digitization became widely apparent in 1993 when [the World Wide Web] broke across the scene. That event brought the clear realization that a new textual condition was at hand and that traditional literary and textual studies had an enormous stake in it. One could now see quite clearly that digitization was both the medium and the message concealed in the crisis that had been developing in literary and cultural studies since the mid-1960s. Why? Because the Web exposes how the technology of archival and bibliographical exchanges can be radically expanded in both spatial and temporal terms. (McGann 2001, 169)

McGann argues that the capacities of digitization and computerized communication illustrate a central tenet of cultural studies, the decentring of the text. The idea of "the text itself," an autonomous and unified work that is an isolated container of meaning, had been under attack long before the emergence of the World Wide Web. But the capacity of HTML coded web sites to offer hyperlinks connecting a document to an infinite number of other texts illustrates in practice the theoretical implausibility of a concept like "the text itself."

McGann's point here is both obvious and breathtaking in its implications: *all texts are already hyperlinked*. McGann (2001, 181) writes, "Every document, every moment in every document, conceals (or reveals) an indeterminate set of interfaces that open into alternate spaces and temporal relations." Texts take on meaning through the myriad ways in which readers connect them up with other sources, use them to make sense of different situations, and draw contrasts with any range of documents in our received cultural tradition. Nothing in this realization (that all texts are already hyperlinked) requires a computer to be shown. "In crucial ways, for instance, a desk strewn with a scholar's materials is far more efficient as a workspace—far more *hypertextual*—than the most powerful workstation, screen-bound, you can buy" (McGann 2001, 185). The idea of a hyperlink is not an alien concept forced upon linear-thinking, page-bound readers; rather, it is a technological development that is successful because it accomplishes something we already do, something, in fact, that is fundamental to the field of English studies. Because we think hypertextually, technology that allows us to make use of that process has tremendous possibilities. Computers fulfil an existing desire much as the printing press did in Gutenberg's time and as the first movie projectors did for the Victorians.

McGann's argument does not stop there, however. Because computer technology can be used to illustrate and further the interests of contemporary criticism, it will be integral to future English studies. In fact, he goes as far as to argue that "The next generation of literary and aesthetic theorists who will most matter are people who will be at least as involved with *making* things as with writing text" (2001, 19). Traditional methods of scholarship, particularly those in which the scholar stood out as unique for having access to remote archival information, make little sense in an age of digitization. Similarly, I would add, it is unlikely that anyone today would go through the effort, for example, to develop by hand a concordance, when any word processor with a "find" function can do in seconds what had once been a time-consuming endeavour.[1]

What is needed is for scholars to apply these digital tools to their areas of expertise. McGann (2001, 169–70) writes,

> . . . it is the literary scholar, the musicologist, the art historian, etc. who have the most intimate understanding of our inherited cultural materials. Hence the importance that traditional scholars gain a theoretical grasp and, perhaps even more important, practical experience in using these new tools and languages. For "theory" in this volatile historical (and historic) situation will have little force or purchase if it isn't grounded in practice.

McGann is not disparaging the skills and expertise of traditional literary scholars, but he is saying that the work of these critics will be more persuasive and have greater impact to the extent that it takes seriously the innovations of computer-assisted criticism. When computers make "practice" this "practical," it is incumbent upon critics to make use of them.[2]

As McGann's argument discredits both the Borg and cyborg perspectives on computerization, so too does William Paulson offer a more nuanced account of the relationship between literature and technology in his recent study *Literary Culture in a World Transformed* (2001). Paulson differs from McGann in that his concern is not primarily with the work of scholarship but with the larger "literary culture" of which academics play a vital but only partial role. Paulson (3–4) defines literary culture as consisting of the "communities, institutions, activities, and attitudes that cluster around" literature, and he includes under this heading, "students and teachers of reading and literature in the schools, people in the book trade, members of book clubs or reading groups, poetry slam aficionados, subscribers to reviews, and ultimately all those who care about reading and writing." The aim of his study is to try and determine the fate of the literary in a culture dominated by the visual and the electronic.

Like McGann, Paulson does not view technology as a kind of alien descending on innocent book readers; rather, he understands specific technological developments to emerge because they address ongoing theoretical and cultural concerns. He writes of our current situation:

> This technological challenge to the continuation of literary culture arrives at a time when many old certainties about the value, centrality, and character of literary study have already been swept aside or at least severely contested. The technological turn away from print and toward electronic textuality, hypermedia, and the audiovisual both complements and radicalises the academic moves away from canonical literature and towards recent, popular, and non-print cultural productions. (Paulson 2001, 9)

Here again we see new technology responded to within a specific historical context. Paulson argues that this technology has become part and parcel of a much larger debate, over the very content of "literature," that is familiar to anyone involved in canon battles or disputes over the significance of popular culture.

Those who lament the decline of print readership miss, according to Paulson (150), the promise that the emergence of new technologies holds for literature: "Now that print has lost much of its dominance among media, it is also losing—and needs to lose—its pretensions to universality, the illusion that it was the natural and obvious way to communicate about almost anything." It is not simply that new technologies will supplement existing print literature but that they will provide different reading experiences to those provided by print. Paulson finds the "advance" of computers to be bringing about the return of a literature closer to that of the oral culture of storytelling. The ease of publication and the possibilities for immediate response via electronic communication makes possible the return of a storytelling in which "the hearer of the tale learns to become its teller" (Paulson 162). However, now the gulf being bridged is not between speaking and listening but between writing and reading. And with the closing of that gap, the distinctions between writers, critics, and readers become less sharp and less significant.

So, as McGann argues that critics need to be "makers" of culture, Paulson celebrates teachers of literature who ask their students to respond creatively to the works they encounter rather than to simply account for the texts they have read. Such assignments allow students to make literature as "equipment for living," to use Kenneth Burke's term (1964). For both critics as well as students, the electronic tools at our disposal will be essential for making literature, whether print or digital, matter in the coming age.

The conclusions reached by McGann and Paulson mark an important advance over the either/or arguments involving technology and literature with which I began this essay. However, literary critics have been by no means alone in tackling these sorts of questions. In journalism, a field within which computer technology has had a far greater impact, there has been another version of the Borg/cyborg argument, only this time with millions of dollars of advertising revenue at stake. While the Internet has driven few print publications into bankruptcy, it is important to note that few "virtual" publications have lived up to the initial hype of the dot.com era. In fact, currently, when the relationship between print and electronic culture is discussed within journalism, the buzz word of choice is "convergence."

In journalism, convergence marks the realization that news stories are often produced for multiple platforms and audiences. A print story may become a broadcast story that may become an online article. Each medium has specific requirements and limitations. Writing journalism without an awareness of the specific media where a story might appear limits the value of an individual's work. To give just one example, "white space" is at a premium in print newspapers. Blank spaces lead to unneeded pages and extra printing costs, so standard paragraphing and layout techniques are designed to use available page space as efficiently as possible. However, online the cost of white space and additional pages is insignificant. What is at a premium online are the eyes of the reader. It is physically more difficult to read a lot of text online, so paragraphing and layout techniques have developed for Internet journalism that make free use of white space in order to ease the strain on readers' eyes. Journalism education is in the process of revamping to acknowledge this new reality and help reporters do the kinds of writing that are being asked for today.

English pedagogy has approached technology much more gingerly. Technologically savvy models of English education are the exception rather than the norm (though clearly some of the souped-up journalism programs rely on students who have been exposed to the kinds of writing traditional English programs specialize in). Nevertheless, we too seem to be in an era of convergence. Few feel (or even wish) that the Internet is going to replace books, but even fewer think of computers as a passing fad. Books and computers will exist alongside each other, in consort with one another. It is up to us whether the relationship between the two is productive or antagonistic.

III

As a practical illustration of the kind of teaching that makes use of a convergence model, I want to turn to a course I teach titled The History and Practice of the Small Press. This class combines instruction in twentieth-century literature with hands-on computer technology experience. While the course responds to the concerns over the acquisition of "computer literacy," it does so by integrating literary movements with technological developments. Students study literature, learn to use desktop publishing software, and produce their own small press publications as a final project.[3] The small press, the independent and not-necessarily-for-profit wing of the publishing industry, has proven to be an ideal means for considering the role of technology in literature instruction. For while it is possible to discuss the small press as a solely literary endeavour, the nature and scope of the small press has been determined largely by the availability of affordable technology. Affordable technology, it is worth noting, is rarely cutting edge. The technology typically used by small presses is widely available or even obsolete. For example, the "mimeo revolution" of the 1950s and 1960s took advantage not only of the mimeograph but also of older offset printing presses that could be bought used. The rise of the Concrete Poetry and Meat Poetry movements during that time depended to a large extent on a producer having access to the means of printing. Though these publications were often of a poor quality by today's aesthetic standards, the formal rawness of work like *The Floating Bear* (edited by Diane di Prima and the-then LeRoi Jones) fit with the attitude of dissent underlying such projects that was realized through the possibility of self-publication.

I teach this class along two mostly parallel but occasionally intersecting tracks. In the first track, we focus on significant moments of noncommercial literary history such as the founding of *Poetry* magazine, Langston Hughes and the politics of patronage, controversies surrounding the publication of Allen Ginsberg's "Howl," Gwendolyn Brooks, Broadside Press, and the Black Arts movement, the rise of 'zine culture, and the coming of hyperfiction. In each of these examples, students not only read and interpret literary texts, but they are asked to consider those texts within a specific cultural context. To take Gwendolyn Brooks' case, what did it mean for her to turn her back on a mainstream publishing contract to work with an African-American-run small press? How can this realignment be read through the poetry she published with Broadside Press?

At the same time that students are tackling these literary and cultural concerns, they also learn the basics of desktop publishing in the second track of the course. For this course, I've made use primarily of Adobe Pagemaker and Photoshop, two readily available computer programs. Pagemaker, though recently replaced by Adobe InDesign, came with a special "build booklet" feature that automatically transposed pages and proved an amazing time saver. Over the course of the semester, students are required to develop a final project, a chapbook or 'zine of their own original material. While creative writers in these classes typically have a reservoir of work to draw upon, other students have put together family or local histories, collections of essays, and even more fanzine-type projects. What really makes this process work is that, at the end of the semester, these publications are sold at a local independent bookstore where we hold a release reception and reading. In the end, computer technology becomes the very means by which students come to participate in literary culture.

It has been a pleasure to find that a focus on the production of a publication has heightened students' attention to language. While I feared initially that students would become immersed in the bells and whistles of desktop publishing, writing for publication has made students more concerned about their language. In fact, the students who are most concerned about the appearance of their publications are also the most careful revisers. Frequent proofreading and last minute editing changes are common when students know their work is going to be published. In this context, students view their choices concerning language use as different in degree, not in kind, from the design decisions they make using computer software.

This combination of creative and critical work with literary and technology instruction is ambitious, but it has been a largely successful undertaking. The class has provided skilled English majors with technological hands-on experience that in some cases has directly related to their post-college careers (in one case for an editor, in another for a student working in advertising). At the very least, students can make use of the ability to self-publish any number of publications.

One ancillary benefit of this course is that it works against the digital gender divide. As many women as men have taken and done well in this course, and we should not underestimate the unique role that literature departments can play here. Women continue to both be underrepresented in technology-intensive fields and to make up a majority of literature majors. English departments can make a unique argument for scarce resources if we can provide technology instruction for students who might not otherwise receive it. Despite these successes, the approach of this class is not without its risks. A high level of technology comes with a high potential for technical disaster, and I've had my share of computer crashes, disk errors, and inexplicable printer snafus (enough, in fact, to make me sometimes nostalgic for the chalkboard).

But a greater problem is that students drawn to the technology of the course sometimes do not have enough prior exposure to literary study to keep up with the literature discussions. It is a class with an interdisciplinary appeal, but a disciplinary focus. I have opted to maintain the disciplinary rigour of the class, but that has been difficult when facing students unfamiliar and uncomfortable with, for example, reading poetry. I see no easy resolution to this particular problem, other than restricting enrollment, as I fear it is endemic to this kind of class, but it is important to expect widely varying exposure to literary study in technology-focused literature courses.

Still, the benefits of the course have far outweighed its drawbacks, and the experience has suggested to me that similar courses are out there waiting to be designed, waiting to be offered, and waiting to bring students into English studies. In the end, I've come away from this experience more convinced than ever that those of us who teach literature do not need to estrange ourselves from the digital revolution. Nor do we need to sacrifice traditional literary study to the god of the screen. But we miss a great opportunity if we do not assert and teach literature's continuing relevance as part of a long history of communications technology. Close reading and attention to detail are both literary skills and part of design aesthetics. In the face of either Borg or cyborg arguments, it is important to argue for the significance of convergence.

Notes

1. A similar argument, though concerning pedagogy, is made by Gregory Jay (1999), who considers what it would take to compile a bibliography of criticism on Louisa May Alcott before and after the digitization of the MLA Bibliography. Jay's point here is that assignments based on compiling information are practically irrelevant today; what is needed are guided research projects that ask students to annotate and evaluate the value of differing sources.
2. In my own experience, I was invited by Stephen Railton, the director of the *Uncle Tom's Cabin and American Culture* multimedia archive, to convert an article I had published in 2000 in *New Literary History*, "Articulating *Uncle Tom's Cabin*," into a multimedia document drawing upon the extensive holdings of that archive. Working with a section of the original article, I created "Grow'd Again: Articulation and the History of Topsy," a multimedia document that is available in that archive. Though *New Literary History* is a respected and widely-held journal, the number of comments, citations and online hits I have received about the multimedia article outnumber those from *NLH* readers ten to one.

3. This type of class is not wholly unique, of course. A number of innovative programs have emerged recently that attempt to integrate technological instruction into literary work. Though the orientation of these programs vary, a short list would include the program in Literary Publishing at Illinois State University; the Ph.D. Minor in Print Culture History at the University of Wisconsin-Madison; a series of specialties within The School of Literature, Communication and Culture at Georgia Tech; and the MA emphasis in publishing and print culture at the University of Minnesota, Duluth.

6

Postcolonial Pedagogical Thresholds: the Imperial Archive and Postgraduate Web Design

Leon Litvack

The Imperial Archive is a web project that forms an integral part of the Literature, Imperialism, Postcolonialism module, taught in the MA in Modern Literary Studies at Queen's University Belfast. Over a period of twelve weeks students examine texts and issues reflecting the influence of the British imperial process on literature of the nineteenth and twentieth centuries. Using colonial discourse and postcolonial theory, the module first examines the British idea of "Empire" and the colonial enterprise in nineteenth-century fiction, and then proceeds to look at twentieth-century texts—some of which "write back" to their predecessors—in an attempt to understand how imperialism continues to affect literary production in Britain's former colonies. The textual pairings include Dickens's *Great Expectations* and Peter Carey's *Jack Maggs* (representing Australia); Charlotte Brontë's *Jane Eyre* and Jean Rhys's *Wide Sargasso Sea* (the Caribbean); and Joyce Cary's *Mister Johnson,* alongside Chinua Achebe's *African Trilogy* (Nigeria). The module is informed theoretically by Ashcroft, Griffiths, and Tiffin's *Post-Colonial Studies Reader* (1995) and *Post-Colonial Studies: the Key Concepts (2000).*

The two-hour seminars are taught in a computer suite, in order to facilitate several teaching and learning activities. Students spend part of their time participating in traditional oral discussion, looking at one work of fiction each week, together with a relevant critical section from the *Post-Colonial Studies Reader*; for example, *Jane Eyre* is studied with an awareness of issues surrounding "Representation and Resistance," while Rushdie's *Midnight's Children* is considered alongside "Postmodernism and Postcolonialism" (Ashcroft et al. 1995, 85–113; 117–47). Printed primary and secondary resources are made available in the library; electronic resources (comprising images, e-texts, sound files, PDFs, digital video, and selected web resources) are delivered through an inhouse virtual learning environment (VLE) known as "Queen's Online" (https://infoserve.qub.ac.uk/home/). Students draw liberally on these materials (built up over the last seven years) for their 5000-word summatively assessed essays (worth 75% of the overall mark). Class discussion is initiated by having each person deliver one PowerPoint presentation (worth 10 per cent) to the rest of the class on a specific fictional text and critical perspective.

The remaining 15 per cent is dedicated to the web project, which represents the most innovative, celebrated, and prominent aspect of the module; it regularly gets 9,000 hits a week from readers around the world. The project's name is partly inspired by Thomas Richards's volume (*The Imperial Archive: Knowledge and the Fantasy of Empire* [1993]) in terms of the accumulation of knowledge, and (in an ironic vein) the author's critique of the control of information for the services of empire. In the context of the World Wide Web, the name conveys the idea that its constituent materials comprise a vast treasure-trove of resources, which have been carefully catalogued, maintained, and scrutinized, to allow for uninhibited, constant access by scholars and enthusiasts across time and space. The first three generations of students (who worked in the years 1996–9) had the hardest job: they established the structure and parameters for the project. The set fictional texts on the syllabus at the time related to literary, political, and cultural expression in six geographical regions: Australia, Canada, the Caribbean, India, Ireland, and Nigeria. These early project contributors were asked to provide overviews of literary and cultural expression in the colonial period; critiques of textual examples; an annotated bibliography (consisting of items they read in the course of their research); and a list of relevant web sites. While these requirements represented a tall order for the students, they had the advantage of establishing a framework that gave a logical structure to each of the geographical subdivisions. The best-known and most visited section is that on Caribbean literature, which the originator dubbed "Christophine," a character in Rhys's *Wide*

Sargasso Sea (Page 2003). In more recent years, students have branched out beyond these geographical boundaries, to examine transnational themes, and other regions not originally covered in the selection of module texts. An interesting example is "The Empire Rides Back," which concerns the world of professional cycling and the building of road networks in the outreaches of empire; the idea developed through the student's interest in cycle racing (Wyer 2003).

Occasionally students have had extraordinary opportunities to engage first-hand with prominent critics, and have incorporated the substance of such meetings into their web projects. An outstanding example is an interview that a pair of students conducted with postcolonial critic Declan Kiberd. Through reflecting on key issues encountered in their studies, the students prepared questions, and recorded Kiberd's answers onto digital audio tape (Faddan and Morrison 1999). This session was then edited using a PC, and the answers to individual questions were uploaded to the Web as digitized audio files. Such an innovative approach gives some idea of the potential that this web project offers postgraduates to create and publish original research.

Such projects as the Kiberd interview require a unique opportunity, a commitment of significant effort, and a high level of technical expertise. All postgraduates in English work within literary and social contexts; they can also understand the effect of theoretical models and critical positions on the development of their discipline. Their development of IT skills, however, does not necessarily extend beyond word processing and the ability to access electronic databases and other information resources. Therefore expectations concerning the contribution they can make to the web project must be informed by the knowledge, understanding, and intellectual skills they acquire throughout their MA, as well as by the skills they can realistically develop in the course of their studies. All MAs in English at Queen's complete a module in research methods, which covers such areas as preparing and presenting a piece of scholarly writing; the use of databases to aid research; discussion of the production and transmission of texts; and the assessment of literary evidence and intentionality.[1] Web authoring ability—which students can only acquire in the Literature, Imperialism, Postcolonialism module—can be considered an additional key skill, which is transferable outside the confines of the discipline of literary studies.

In the early incarnations of the module, students used a free trial version of Softquad's HoTMetaL, designed for Windows 3.1. In the mid-1990s, this authoring package was popular with both amateur and professional web developers, because of its flexibility, ease of use, and WYSIWYG ("what you see is what you get") interface. Despite its advantages, the cost of installing later full versions of the software (designed for a Windows 95/98 platform) on PCs within the university was prohibitive; for this reason the project abandoned HoTMetaL in 1999. The only viable alternatives at the time were Microsoft Word and FrontPage, both of which formed part of the university's Microsoft Select agreement, and were available on PCs in open access areas throughout the institution. While it would initially seem advantageous to use Word for HTML editing (because little extra tuition is required), its problems are well known: for example, Word introduces extraneous HTML code that is required to format and display documents in Word, but is not needed to display the HTML file. This problem can be overcome by employing an HTML filter or converter; however, this strategy does not allow the student to learn about HTML code, and perpetuates the problem of writing "bad" HTML. FrontPage also has problems as a web editor; for example, it relies on a Microsoft server technology to make all its features accessible; also, bullets and tables are not readily formatted. Despite the acknowledged difficulties, these imperfect tools were used for the project between 1999 and 2001, because they were the only ones the university made universally available. By 2002 the university moved to adopt Macromedia Dreamweaver as its web authoring package of choice. It had great advantages over its predecessors, including its easy-to-use templates and cascading style sheets (CSS), and its facility for administering sites to which multiple authors/developers contribute. Though the learning curve for Dreamweaver is much steeper than for Word or FrontPage, the potential rewards are greater.

The web project combines traditional and innovative methodologies to produce an exciting research resource; it also raises interesting issues concerning assessment as part of a degree in literary studies. After some experimentation, and consultation with the external examiners, it was decided that the School's standard marking criteria could be employed,[2] but adjusted to take into account the peculiar features of the Web. The six criteria are relevance; knowledge; analysis; argument and structure; originality; and presentation. These criteria provide teachers of English with readily identifiable touchstones in essays of the type which students of English have been accustomed to write in their undergraduate and postgraduate careers. These established points also provide students with a clear idea of what is expected, and ensure that the goals are achievable. For example, in the context of an essay, in order to attain a first-class mark, a student's response must be directly relevant to the question, and must consider the implications, assumptions, and nuances of the question. It must demonstrate an excellent degree of knowledge in breadth and range of reading, and must show a very good analytical treatment of the evidence, resulting in a clear synthesis. The answer must also display a coherence and structure. In order to satisfy the criterion of originality, it must be distinctive, displaying independence of thought and approach. Finally, it must be well written, with standard spelling and syntax, composed in a readable style, and with appropriate documentation. The majority of contributions take the form of short essays of the students' own design, which feature a coherent argument, like the work they are accustomed to doing for their other modules; they also feature relevant web links, and a bibliography of printed sources consulted. They are asked to produce a total of 3,000 words of text; often this requirement results in the construction of up to three web pages, but occasionally students opt for two slightly longer pieces. It is essential that they display an awareness of the potential of the Web for enhancing their arguments beyond the written word.

For most of them, this process involves the inclusion of appropriate images. The more adventurous students will take photos themselves (Burke 1999); most, however, will make do with easily obtainable images already in the public domain. When work on the pages is complete, they are checked, uploaded to the Web, then marked by two internal examiners and the external examiner. By the time of the examiners' meeting, all markers are expected to have reviewed the material online, in order to appreciate the nature of the medium, and to observe how students' contributions accord with the overall conception of the site. Mindful of the six criteria, they assess individual contributions as a "package," and assign to each student a single agreed mark, based on the 17-point University mark scheme of conceptual equivalents and percentage grades, ranging from a "high/excellent first" at 90 per cent, through a "definite/solid II.1" at

65 per cent, down to "nothing of merit" at zero. This mark is then converted to a score out of fifteen, to arrive at the final grade. If any errors or contentious points are found in the pages, these are corrected before the final mark for the module is released. Each student-generated page also carries a statement, which reads:

> This project was completed under the direction of Dr Leon Litvack as a requirement for the MA degree in Modern Literary Studies in the School of English at the Queen's University of Belfast. The site is evolving and will include contributions from future generations of MA students on other writers and themes.

This imprimatur acts as a form of "quality assurance," informing the reader that the pages form part of a larger student-led project and university degree programme, and that they have been scrutinized by members of academic staff; thus "quality control" is assured.

For the students themselves, the experience has proved a rewarding one. This is indicated not only through the enthusiastic comments observed in the questionnaires, but—more importantly—in the use students are able to make of their contributions to the site after they leave the university. All pages carry "mailto" links, which most keep updated to reflect changes in their e-mail addresses. They clearly enjoy receiving feedback from readers, and in some cases (such as the pages on the Caribbean, India, and Nigeria), the authors have engaged in debate long after graduation: of particular note is one student who had a five-year debate with a reader concerning her views on the "Indian Mutiny" (Fallon 1997). The enhancement of transferable skills through web authoring has proved to be a point of discussion in professional contexts: those who subsequently applied for positions in teaching and the media were asked to reflect on their experience of *The Imperial Archive* in job interviews. In three other cases, the work on the Web has led graduates to undertake further IT training, in the form of an MSc in computer science for nonspecialists: all three have confirmed that had they not been exposed to the web authoring component in their MA, it would not have occurred to them to look towards careers in IT. Two students have gone on to doctoral work in areas that they were first able to explore in web projects: one on the fiction of Peter Carey, and another on the image of the Tinker in Irish literature and culture.

The success and reputation of *The Imperial Archive* as a reliable research resource are confirmed by the number of sites that have requested links to it, or which have reproduced material from it. Of particular note is George Landow's *Contemporary Postcolonial and Postimperial Literature in English* site, hosted at the National University of Singapore (http://www.postcolonialweb.org/). Landow's site developed along similar lines to his long-established *Victorian Web* (http://www.victorianweb.org/), which originated at Brown University in 1995.[3] Contributors to the "PoCo Web" include established scholars, a host of undergraduates from Brown, and a small group of individuals from other institutions. Landow asked if several contributions on Peter Carey from *The Imperial Archive* (Dunlop 2003) could be duplicated on his site, and permission was readily granted. The pages were reproduced with due acknowledgement of their source, and have helped to publicize *The Imperial Archive* more widely.

There have been other instances of cooperation in electronic media. For example, links to several pages now appear in two resources published by ProQuest Information and Learning (formerly Chadwyck-Healey Ltd). The first, *Literature Online*, features Chadwyck-Healey's full-text databases in English and American Literature, and is available in hundreds of libraries and academic institutions worldwide. The second, *ProQuest Learning: Literature*, is designed to support the teaching and study of English literature at A Level, AS Level, national qualifications in Scotland, and for the International Baccalaureate. It offers students and teachers access to a large archive of primary and secondary materials relevant to the texts, authors, and topics set by the exam boards in the UK. ProQuest aims to provide access to the most informative and accessible free web resources currently available on key authors and works; pages that have been reproduced from the *Imperial Archive* cover such writers as Brian Friel, Chinua Achebe, Jean Rhys, and Douglas Coupland, and include such topics as biography, history, and language, as well as colonial and postcolonial contexts (Morrison 1998; Faddan 1998; Page 1997; Slattery 1998; and Martin 1998). These links have given wider publicity to the site, and have made greater numbers aware of its value and usefulness as an educational resource. Finally, a new area of recognition of the *Archive*'s importance is print media. Requests have been received from editors of anthologies for inclusion of essays in published volumes; for example, the page on magic realism in *Midnight's Children*

(Stewart 1999) will appear in a collection of scholarly articles on Rushdie.[4]

There are plans for future development of the project. Taken as a whole, the pages seem, at present, somewhat eclectic in design, and the project has outgrown its original parameters. While the absence of a consistent format was acceptable when the project was in its infancy, if it is now to embrace fully the role of a professional, scholarly research resource, more uniformity is required, with better navigation through the site. The idea of writing short essay-style pages will be retained, because this is not only what the students are accustomed to, but it also sharpens their powers of argument and analysis; the format has significant appeal to readers as well, because the material goes far beyond introductory commentary. Another difficulty is that there is at present no effective way of accommodating those students who wish to write comparative pieces, that cross geographical boundaries, because such contributions cause problems with classification (that is, should the page belong in one geographical region or another?). The project must therefore carefully negotiate between individual aspirations and the project's overarching vision and structure. There is also a need to provide a forum for exploring and understanding postcolonial theory as a set of coherent ideas;[5] this development would assist the students themselves to participate in a more general debate, and move beyond an author- or region-centred approach. Once the structure has been streamlined, the project will introduce Macromedia Contribute as the student authoring package of choice. This application, which is as easy to use as Microsoft Word, allows for the use of templates and cascading style sheets: all the students need to do is to add content into editable regions of the page. Contribute does not require a knowledge of HTML, and so is much easier to learn than Dreamweaver; the time saved could be better spent on dealing with the issues that lie at the heart of the module. Because the project leader retains control of page design, code, and permissions—these cannot be changed by the Contribute user—integrity of the web site is maintained.

The Imperial Archive has grown to become a respected and relevant research facility in postcolonial studies in English. From its rather humble beginnings as an inhouse postgraduate project, designed to highlight innovative assessment, it has become the most frequently consulted set of pages on the Queen's English web site. Its reputation has been achieved with little active promotion; instead it has relied upon its readers to publicize its activities and resources. Now that it is a well-recognized site, it needs more careful organization and forward planning of content to ensure a continued relevance and augmentation. This can be achieved with manageable effort on the part of the project leader. It is hoped that this outline of conception, procedures, and content will assist others in developing innovative student-led projects that do not involve significant commitment of either human or material resources, and can be adapted to a variety of pedagogical circumstances.

Notes

1. See http://www.qub.ac.uk/en/teaching/postgraduate/modernma.htm#110ENG760.
2. See http://www.qub.ac.uk/en/resources/marking.htm.
3. See http://www.victorianweb.org/misc/credits.HTML.
4. This collection will be published in India, under the editorship of Dr Mohit K. Ray of Burdwan University.
5. See Landow's attempt to provide an overview of postcolonial theory at http://www.postcolonialweb.org/poldiscourse/discourseov.HTML.

Part 3

Virtual Teaching and Learning Environments

7
All Aboard Blackboard

Lisa Botshon

The tools I currently use in my English classroom are quite primitive: mostly lecture, chalk and board, and print materials. Aside from a few DVDs and Web-based assignments, the technology I use for teaching is fairly archaic. My nineteenth-century predecessors would find my classroom shockingly familiar, right down to the chalk dust. One might imagine that I am some sort of technophobe, then, resistant to incorporating new forms of technology into my pedagogy. But this is far from the truth; like many colleagues, I have sought to embrace technologies that might help to make my classroom a better space for learning.

So what might account for my reliance on primitive materials? The fact is, we all still use them. For a significant portion of the student body, these old tools do promote learning. We have thoughtful and provocative content; lively visuals, even if they are comprised mostly of ourselves and overheads or PowerPoint images; the sounds of our discussions and lectures; and a variety of methods to combine and manipulate them, with which we hope to reach a variety of learners. For decades educators have been discussing the ways in which we might employ new technologies in our institutions of higher learning. Through these new tools, universities hope to make education more accessible, increase learning potential, and, not least, make a profit: all of which are valid aims. And the recent boom in educational technology suggests that educators believe in the tremendous potential of technology as a learning tool. What happens, though, when the technology does not fulfill its potential? What happens when universities subscribe to expensive new tools that fall flat rather than enhancing the old techniques or moving us to new levels of pedagogical success? This essay examines the challenges associated with institutional decisions to acquire educational technologies and looks especially at the problems of one of the most widely used forms of courseware, Blackboard.

Teaching technology

Long before the virtual campus arose as a new form of university access, my university, the University of Maine at Augusta, sought to address the public higher education needs of those who could not attend a traditional campus and embraced alternative forms of distance education. For this purpose, the interactive television (ITV) system was developed in the 1980s and is still used to this day. This technology replicates the teaching paradigms of the traditional classroom: instructors lecture in front of a camera and, usually, live students, using traditional teaching tools such as video, audio, and overheads interspersed with computer images and sounds. The classes are broadcast via a fibre-optic network to students at sites and centres all over the state of Maine: all have access to a toll-free phone so that they may call in to answer or ask questions, or add their commentary. In this way, the university has made a significant attempt to reach students who may not otherwise be able to attend university classes.

While my paradigmatic nineteenth-century teacher might be baffled by the three ceiling-mounted cameras, the television screens, and the technician in the "producer's" booth in the back of the typical ITV classroom, he or she would be quite comfortable conveying information to students in the ways he or she always has. After all, the modes of pedagogy employed in the ITV situation are not really that different from those used in a traditional classroom.

However, this technology has limitations that traditional classrooms do not. Though ITV is not entirely a passive mode of course delivery, students are not encouraged to participate in much immediate active learning, except via telephone, which is optional for most courses. (It has to be—many ITV courses enroll over sixty students; required courses without enrollment caps, such as the American History survey course, typically contain well over 100.) Moreover, there are noticeable time lapses between an instructor's question and a distance student's answer; shy students are even less likely to call a class of strangers than raise their hands in a traditional classroom; and managing large numbers of bodies in different spaces at one time is incredibly time consuming for the instructor. In addition, the large number of essays and exams that are mailed back and forth are too frequently missing in action, ending up in other instructors' piles or just delayed by the post: on my campus, for example, student evaluations of ITV courses are usually one or two points below those that are taught in traditional classrooms—even with the same instructors, texts, materials, and assignments. It is widely recognized that student satisfaction is typically lower in these courses than in parallel courses taught in traditional classroom spaces. I can't imagine how it would be otherwise: distance students who enroll in ITV courses are often sitting alone in a room in a high school or community centre staring at a television screen where a talking head is sometimes interspersed with other images, like text or data. They lack the energy and enthusiasm generated by groups of people who are in the same place at the same time. And they may not be learning the material as well as their traditional counterparts; based on my assessments in my Introduction to Literature course, which are mostly via essays, ITV students receive slightly lower grades on average than their peers in traditional classes.

Finally, ITV classrooms are quite expensive to run. Each course requires not only the services of a professor, but also a room equipped with a tremendous amount of machinery, including the aforementioned cameras, televisions, and computers. There is also the "producer's booth," in which a technician is employed to run the cameras, field the students' calls, and mount any extra images or sound. Not least, there are the technicians and administrators who keep the whole enterprise afloat: from the maintenance of the satellite transmissions to the handling of exams and papers—in short, it is a labour- and machine-intensive endeavour.

Technology in transition

In the late 1990s the academy was just beginning to grapple with the concept of delivering courses entirely online. For UMA, online courses appeared to be a cheaper, more effective way of reaching out to students at a distance than the clunky technology of ITV. Students would not have to leave home at all if they had access to a computer and a modem, and our "nontraditional" working/parenting students would benefit from the asynchronous nature of the web-based course, logging in whenever they had time. Additionally, the university would not need to employ so many technicians and administrators to run this enterprise: all we would need was some software and a few people to ensure that it worked. The professor would handle most of the rest of the labour. The University of Maine System chose Blackboard, a commercial brand of courseware, to implement its online offerings.

At first glance, Blackboard appeared to solve many of the challenges facing distance education faculty, administrators, and staff. Staff were too few to design specialty programs for individual courses. Administrators wanted to appeal to place-bound students, but without the astronomical costs associated with technology like satellite television. And faculty wanted technology that would be useful and accessible to both themselves and their students. Blackboard required few technical skills from any of its users, required (relatively) little maintenance on the part of the tech staff, and was (relatively) affordable.

Blackboard is one of the most popular web-based software programs that provide online class rosters, syllabi, assignments, a drop box for papers, gradebooks and vehicles for discussions, quizzes, and exams. According to *Chronicle* reporter Florence Olsen (2001), it grew out of the entrepreneurial programming efforts of seven undergraduates at Cornell University, who helped a business professor build a course web site. Tellingly, according to Olsen, Blackboard was not "originally meant to be a pedagogical tool." Nonetheless, Blackboard has sent highly persuasive representatives to academic administrators and technicians in recent years, and their smooth marketing has rendered their product among the leading commercial coursewares in use. Blackboard trainers who came to our school told of students participating in online conversations about scholarly materials that were so enthralling that these discussions continued well after the courses had ended. And, unlike regular classroom experiences, they didn't have to end! Students could take their excitement about literature or sociology out of the realm of the university and just keep it going on their own. All they needed, after all, was Web access. It sounded like a wonderful way to teach: students would be constantly plugged in and interacting, and teachers could finally bypass the clunky technology of ITV and stage meaningful learning experiences in everyone's living rooms and offices.

An initial foray

In 2000, when I was presented with the opportunity to teach my Introduction to Literature course online, I was enthusiastic about the possibilities such a format engendered. The staff of the University College, the branch of the University of Maine System that now handles all of our distance education, helpfully directed me how to transfer my syllabus, weekly discussion questions, and essay assignments to the software and, indeed, as the Blackboard reps had promised, it was fairly easy for me to learn how to manipulate all the parts of Blackboard that I needed. And, as with ITV, my course looked quite familiar: instead of lecturing and posing verbal questions, I could post background information about the texts we read and then send written questions for the students to discuss. Students would debate the same issues as in my traditional classroom. Essays retained their format. I did discover that it would be necessary to add extra information to my course, like directions on how the course would run online, and how students could access technicians to help them with the software. Overall, however, it seemed like a fairly painless—and promising—transformation.

Over the spring of 2001, though, as I attempted to teach twenty online students the same sorts of materials I had been successfully teaching my other Introduction to Literature classes for three years, I realized that the course was not going well. The first tip-off was my discovery that within a month of making initial contact, exactly half the class had dropped out. The second was that two of the remaining students began to "flame" me (or send derogatory comments to me on a regular basis). And the third was the emergence of web-based plagiarism. What could I have been doing so wrong so as to engender such radical student abjection and misbehaviour? I attempted to make reparations with more direction, a lot of midnight interventions, and a typical humble-teacher response: "Here's the discussion board. You tell me what you want—even anonymously." But this did little to improve the quality of the class, and we limped along until the end of the semester, when the ten remaining students and I heaved a collective (our first) sigh of relief.

Then I went back over all the discussions and e-mails and final evaluations to see what I could uncover about, and hopefully learn from, the disaster that had been my first online course. What I found was discouraging. Several students had dropped out early in the semester because, as easy to use as the software appeared to be, they were completely flummoxed by it. One of them had never even e-mailed before and, despite my many phone calls and pleas to visit my office (he didn't live that far away), it took him a month to figure it out. By this time, it was much too late for him to start learning the ins and outs of Blackboard, not to mention to catch up on the readings and assignments. A couple of students lacked home or office access to a computer and found it difficult to log in regularly from local libraries or parents' or friends' houses. A few students admitted that they thought an online course would be "easier" than ITV or traditional classrooms, and dropped out when it turned out to require a tremendous amount of writing time. Evidently, a majority of these students had signed up for the course without a clear picture of prerequisite skills or resources. This problem might be solved with better marketing and initial questionnaires or checklists to ensure better understanding of the course before registration. But what of the flamers and plagiarizers and the other students who just hung in there joylessly, hoping to get through a bad semester?

The plagiarizers, it turned out, had no idea that they were violating any sort of code of ethics. When I ran one clearly copied essay through a plagiarism detection program on the Web, I discovered that the student had worked very hard, downloading at least ten other people's perspectives of Ibsen's *The Doll's House* and then weaving them together. She thought that this was the correct way to write an essay. The fact that so much academic work (both good and bad) is now easily available online suggests that our students need more guidance in how to evaluate and handle it. Plagiarism is a phenomenon that is not unique to the online class—although one might argue that virtual classes could inspire virtual work, web-based plagiarism was beginning (and continues) to plague the academy at large. My Ibsen plagiarist inspired a solemn reexplanation of essay writing and source citing posted prominently in the next week's class notes.

The flamers were a different matter. Sometimes students who send flaming remarks are just bullies, analogous to the occasional loudmouth that can crop up in a traditional classroom. But in this case, I was fairly sure that these students did not mean to sound derogatory; their messages were off tone because they were frustrated and they were not yet good enough writers to know how to best communicate with their professor solely through e-mails. Neither of these students had taken an online course before and both of them were used to talking to their professors in person when they had a problem. When I got the flamers on the telephone, we were able to have a much more productive conversation and the flaming ceased. Of course, as Rena Palloff and Keith Pratt (2001, 113) argue in their book *Lessons from the Cyberspace Classroom,* some students just do not learn well online.

If you add up the list of problems my online class incited here, it is tempting to view them merely as the challenges of a new format and to dispense with them piecemeal: first make sure all the students know how to use a computer and e-mail; then make sure they have regular access to a computer with web access; conduct workshops on proper citation and how to use web sources; finally, explain how e-mails to fellow students and professors should be appropriate in tone and content. Surely these directions would have improved certain aspects of my course.

However, a larger problem superseded, and perhaps provoked, all the others: my online course was a pale, flat version of its traditional form and I could not employ even the kinds of teaching and learning styles available to my nineteenth-century predecessor. As philosopher and technocritic Andrew Feenberg (1999, 29) writes, faculty must ask, "How . . . can one duplicate on an electronic network the learning experience of a highly interactive classroom? How can one reproduce the wealth of informal human contacts that add so much to education on campus? And how can the intense moments of personal interaction that mark our memories and our lives ever occur in a sterile electronic environment experienced in the isolation of the home?"

Lacking sound (discussion), visuals (beyond the texts and a few images here and there), and personal interaction, Blackboard's technology actually limited the kind of teaching and learning that could take place; the only student who could possibly excel here was someone who was not only self-motivated, but also already a good reader and writer. Feenberg (1999, 30) reiterates the limitations of this medium: "The online environment is essentially a space for written interaction . . . Electronic networks should be appropriated by educational institutions with this in mind, and not turned into poor copies of the face-to-face classroom they can never adequately reproduce." I have no doubt that had the class been made up of upper-level

English majors or, indeed, my colleagues, that the work of Introduction to Literature online would have been much more productive. But, then, advanced readers and writers don't require a course like Introduction to Literature.

One might suggest that perhaps this particular course is not well suited to Blackboard-type launching. But these larger problems persist in the wider arena of the academy. For example, Dan Carnevale (2003) reports that, according to the results of a recent survey taken in the University of Wisconsin System, a system that used Blackboard as its primary form of courseware, "faculty members find course-management systems time-consuming and inflexible, and students find them difficult to use. Some faculty members . . . reported that their students actively discourage the use of course-management systems."

What is to be done?

It is critical that we altogether rethink the way we teach when we employ new technologies. In retrospect, merely transferring my traditional course materials to an online format was destined to fail. But even after my initial foray and analysis was complete, I would not have come up with a solid alternative on my own. Blackboard does not lend itself to innovative pedagogy, and the ways in which we are "trained" to use it has more to do with learning the functions of the courseware and little to do with teaching techniques.

Palloff and Pratt (2001, 152) argue that in the cyber classroom, course development needs to focus on interactivity, *not* content, which is a huge difference from traditional classroom technique. According to these authors, professors need to learn how to decentre themselves and act as directors rather than founts of information. Students, who often expect that the teacher is the only person they can learn from, have to learn how to work and learn more collaboratively; they need to be "oriented to their new role and the ways in which learning occurs online" (2001, 153). Moreover, they assert, "learning through the use of technology takes more than mastering a software program or feeling comfortable with the hardware being used. Students in online learning situations need to come to an awareness that learning through the use of technology significantly affects the learning process itself" (2001, 108). Blackboard does not easily facilitate these processes as it only replicates in virtual form the tools that we use in the classroom. If we are to truly commit ourselves to employing new technologies to improve access and the quality of learning in our universities, then we must take up new tactics.

First, we need to release the stranglehold Blackboard and other commercial products have on so many universities. John Unsworth, writing for the *Chronicle* in January 2004, explains that Blackboard became indispensable in the late 1990s and early years of the new millennium because universities realized that they were unable to meet their own specialized information technology needs with homegrown software. Commercial products appeared to be the only (affordable) alternative. However, more recently, open-source efforts that allow consortia of schools to share different kinds of software and adapt them to their individualized needs have weakened the grip of commercial products. Even Blackboard has seen the writing on the wall and has begun to relax its restrictions so that colleges can create their own software programs and integrate them. What's important about this movement, according to Unsworth (2004), is that universities will get "greater portability of content, greater flexibility in choosing and assembling elements of a learning-management system, and a shift in the balance of power between the client (the university) and the software vendor, in favor of the client." More pointedly, universities will be able to expand the kinds of software they use for teaching and be able to experiment more with more effective pedagogical tools.

Second, we need to involve professors more fully in information technology decisions. Traditionally, administrators, not faculty members, have made the largest decisions about educational computing. And, as Andrew Feenberg (1999, 28) has argued, "For too many administrators, the big issues are not educational. What interests them are the fiscal implications of online learning. Administrators hope to use new technology to finesse the coming crisis in higher-education spending, and to accommodate exploding enrollments of young people and returning students." But professors *should* be at the forefront of educational technology development and adoption; they are the ones who are interacting with students; and they are the ones attempting to foster educational communities.

In a related fashion, universities must become more imaginative about creating educational technology. Obviously, faculty cannot do this on their own; most of them are not trained in code, and, as we all know, there are myriad other duties, including teaching and research, in which they are involved. Moreover, creating innovative pedagogical solutions requires much more knowledge and creativity than merely learning HTML or Photoshop. Corinne Laverty et al. (2003) have noted that faculty members are usually left with little technological support other than how to use the mechanics of a given form of technology. Typically, there is little guidance on how technology relates to course goals or enhances education. Laverty's group has found that a more successful method of developing educational technology is the formation of a team of collaborators with diverse expertise, which may be comprised of instructors (subject expertise), librarians (information resources), instructional designers (pedagogical tools and learning outcomes), and technical support personnel (hardware and software). According to these authors, "The goal of the team is to enhance the learning environment in student courses through identification of teaching and/or learning challenges" (2003, 20). Furthermore, in his article "Bring in the Geeks," Peter Schilling (2003a, 13) makes a compelling case for including nontraditional cyber experts in educational technology development. "[W]e must bring the database programmers, graphic artists, and game developers" to the students and faculty, he says. People with these skills can help universities develop new ways of creating and using knowledge.

Not least, we must ensure that the students themselves have adequate access to and are trained and supported in the new forms of technology we employ in our universities. When we still have students who are completely unfamiliar with e-mail and the Internet, and when a great percentage of our students who do use the Web (at least here in the predominantly rural state of Maine) have access only to slow dial-up modems, we must work harder to guarantee that they can do the work we ask of them.

Conclusions

Universities have rarely been at the forefront of developing new technologies. But this does not mean that we—faculty, students, administrators—are not ready to take on this new world. Even learning the most traditional of subjects, like English, can be positively transformed. Randy Bass (1998, 12), an early proponent of employing new forms of technology in his American Studies courses, perceives that there are several ways in which new technologies can play a role in increasing learning: "[T]hey allow students to have direct access to the growing distribution of cultural knowledge across diverse resources; and they provide means for the distribution of responsibility for making knowledge in the classroom, by giving students media through which to construct and share their ideas about these materials in a whole range of public learning contexts." Bass' guide to using technology to teach American culture lists a number of interesting educational examples. For instance, at Bowling Green State University, a class studying the American 1890s developed a web site that provides links to the era's social, political, and cultural events. The students themselves conducted the research about these subjects and also created their own photographs and links to other relevant sites. Undergraduates at Georgetown University have transcribed and analysed documents for the Jesuit Plantation Project, mounting primary sources and helpful maps and links on the Web (Bass 1998, 258, 261).

Bass's constructive ideas rely on technologies our universities already have, insisting on new ways that teachers and students might employ these tools for better learning results. But these concepts are just the tip of the iceberg. Peter Schilling (2003b), in his article on the evolution of educational technology, foresees far greater changes ahead. He envisions "tools to track what and how students learn and then deliver the appropriate content to each," something that our nineteenth-century predecessors could never accomplish, at least not on the scale and with the effectiveness of the new forms of adaptive learning courseware (2004). He, like Palloff and Pratt (2001), sees the roles of professors changing, becoming more about managing and directing student projects and less about providing content. Most transformative is his vision of 3-D multiuser worlds, akin to games already in existence like Sim City, "models that evolve in sync with the real world . . . Using GIS/GPS, automated data collectors, 3D imaging, and other technologies . . . the post-organic learning world would grow and evolve as the actual world does. It can also model potential futures and pasts" (Schilling 2003b). Hence students studying Renaissance England could gather historical, linguistic, literary, and geographic data and use these tools to recreate the conditions of the travelling theatre troupes of the past. It is hard to imagine a greater departure from my flattened Blackboard literature class.

Clearly, information technology is the wave of our educational future. But we must be savvy about how and why we are using it. We must strive to ensure that fiscal and political concerns do not supercede educational ones. And we must develop and employ technology that will truly aid our students' learning. To do so effectively, we should open the coffers, think creatively, and work collaboratively. And when the sirens of products like Blackboard beckon, we must evaluate their allure against our larger goals.

8
American Cultural Studies and e-Teaching Internationally

Dorothea Fischer-Hornung and Wolfgang Holtkamp

Introduction: consequences of the communication age

Society and education exist in a dynamic, mutually determined, and determining relationship. As a result, paradigm shifts in society lead to similar changes in education and vice versa. Consequently, whether we label the current period the age of information or communication, the unparalleled social changes attending the age require a fundamental redesign of social and educational institutions.

The rise of modern technologies presents us with the challenge of transforming unrelated, disparate, and isolated pieces of information that are increasingly easy to access into networked systems of knowledge that cannot only be applied immediately but changed and modified with great flexibility as the need arises: "With so much information available, we need people who can synthesise meaning from large bodies of diverse knowledge; and with the arrival of the communication age, we begin to realise that this meaning-making activity is a highly collaborative process" (King 1998, 365–6). When learning, thinking, and working are no longer solitary activities, as King indicates, then traditional notions of teaching must be redesigned throughout our educational institutions in order to meet the challenges of the communication age—teaching the humanities at our universities cannot be an exception.

This essay explores the implications of the explosion of accessible knowledge and the challenges of systematized e-teaching of American cultural studies on a national and international scale via the Internet. It will introduce American Cultural Studies Onweb, a project that since 2001 has designed, developed, and implemented a series of online courses taught at the Universities of Heidelberg and Stuttgart, Germany. This project not only addresses the challenges of the new era of communication, the changed form of media and information culture, but also the fundamental transformations that marked the transition from American studies to American cultural studies, reflecting changes in American society since the 1960s. Finally, we will outline some specific conclusions about electronic teaching in the humanities based on the experience garnered during the initial phase of the project.

From American studies to American cultural studies

Over the past several decades the discipline of American cultural studies has evolved significantly, reflecting the important influences of, for example, women's studies, ethnic studies, as well as postmodern and postcolonial studies. In his book *The New American Studies*, John Carlos Rowe (2002, 51) points out that "[i]n response to concepts of American identity shaped by Western patriarchy and Eurocentric models for social organization, more recent critical approaches have focused on the many cultures that have been marginalized by traditional American Studies or subordinated to an overarching nationalist mythology." Whereas traditional American studies relied on the model of a single dominant culture, more recent approaches focus on the differences among the many cultures constituting the United States of America. Rowe (2002, 53) concludes that "the dominance of the United States according to the nationalist paradigm has often led to the neglect of other nations of the western hemisphere." Consequently, the "new American Studies tries to work as a genuinely comparatist discipline that will respect the many different social systems and cultural affiliations of the Americas." The concern about cultural differences is analysed in the context of their mutual contact and interaction. Scholars who share this vision are not limited to the United States alone and recently the American Studies Association has specifically addressed and welcomed the international exchange of scholarly work "for the benefit of both US and non-US scholars and in recognition of the very different purposes, interests, and institutional configurations American studies may have around the globe" (Rowe 2002, 56). This development is also reflected in the founding of various internationally-based associations devoted to various aspects of American cultural studies, for example, CAAR, Collegium for African American Research (Tenerife 1995), MESEA, the Society for Multiethnic Studies: Europe and the Americas (Heidelberg 1998), and IASA, the International American Studies Association (Leiden 2003).

The fundamental reconsiderations of what constitutes American studies of necessity must, on the one hand, be reflected in the content of our curriculum. The advances in the communication age, on the other hand, must also find their way into the methodology of how these contents can be taught most effectively. Like Rowe, we believe that the new and extended vision of American cultural studies often stands in contrast to a university that stresses the traditional transmission of knowledge and certainly the curriculum often does not reflect the radical changes in the technological environment of the students.

Fundamental changes in the way universities educate their students—both in form and content—are necessary to fully make the transition from American studies to contemporary American cultural studies. As a consequence the classroom will have to be transformed "from the traditional scene of instruction . . . into a joint venture involving many scholars, including our students as active researchers" (Rowe 2002, 61). For Rowe, earlier examples of active–passive models like team-teaching or coordinated classes find their modern answers in "alternative learning situations offered by the Internet, distance learning, and other electronic means of instruction. Electronic MUDs (multiuser dimensions) and MOOs (multiobject orientations), virtual conferences, and hypertext databases should be used as more than mere tools in traditional classroom education and conventional research" (2002, 61). We have taken up this challenge and, like many others in the growing group of e-teachers, we believe that the new electronic information technologies should reflect the fundamental paradigm shift in our concept of what education and knowledge means generally, and in the humanities specifically.

Pedagogies for e-teaching

Education involves the creation of knowledge through dialogue and interaction; the Internet offers an excellent environment for both. Today many people use the Internet for shopping, communication, and general leisure. Therefore, according to Martin Weller (2002, 10), "[i]t would perhaps be foolish to assume that in such a climate people will not expect to have their education via the same means." Further, Weller (2002, 33) predicts that the next twenty years will bring about greater change in education "than has been seen since the foundation of universities removed knowledge from the power of the church." Weller (2002, 30) does not expect academics to become experts in software development but they must learn how to avail themselves of the potential of new technologies: "just as they are expected to be able to use the library effectively, they should be able to use the new information tools." Moreover, he points out that to close our eyes to this development is to fail to meet the challenge of the twenty-first century: "The way to ensure the quality of online education is for educators to become involved with the process, not to refuse to engage with it" (2002, 30).

Computer-based training (CBT) and computer-assisted learning (CAL) gained popularity with the introduction of personal computers in the 1980s and saw explosive growth during the 1990s, with widening acceptance of multimedia approaches. Often CD-RoMs provided additional support to traditional teaching methods, and they were employed in both face-to-face and distance education. While they were especially suited for simulations in which students could experiment with variables and bridge the gap between theory and reality, they were not influential overall upon general education (Weller 2002, 58).

The growth of the World Wide Web, however, has stimulated a new interest in CAL. The likely increase in Internet accessibility, improvements in bandwidth and general penetration into all areas of society are good indicators for the inevitable spread of electronic education (Weller 2002, 63). Any online course by its very nature faces the challenge of combining new technologies with a changed didactic approach. Constructivism offers a useful pedagogical framework for approaching online teaching. According to the constructivist approach, "learners construct their own knowledge, based on their experience and relationship with concepts. Each learner therefore has a unique representation of the knowledge formed by constructing his or her own solutions and interpretations to problems and ideas" (Weller 2002, 65). Influenced by the works of psychologist Jerome Bruner, constructivism has become a broad term to cover a general approach, emphasizing both the active nature of a student's own ability within a carefully constructed learning environment and the process potentially inherent in such a learning approach. It is more a learning theory than a teaching theory. Some characteristics of constructivist courses outlined by Weller (2002, 146–7) include:

- an emphasis on student interaction;
- assessment that focuses on process and student interpretation;
- an emphasis on students' own experiences and their understanding of concepts;
- use of techniques such as dialogue and collaborative working;
- learning as a social activity;
- the educator in the role of facilitator and possibly mentor.

Weller (2002, 147) divides the broad range of electronic courses into four categories: "high technology-didactic," "low technology-didactic," "low technology-constructivist," and "high technology-constructivist." Our own ACS-onweb project is based on the "low technology-constructivist" approach (Weller 2002, 149), utilizing relatively simple technology in an e-learning platform environment, such as a web site containing text, links, images, and computer-mediated means for synchronous or asynchronous communication (e-mail, instant messaging, and bulletin boards). Much of the learning is organized around communication, dialogue, and interaction, a methodology especially suited to online courses in the humanities that require a good deal of discussion.

If the Internet is used as resource, the amount of course material on the web site can be minimized. Nevertheless, the online courses require a carefully structured framework that fosters interactions and independent research using the resources available on the Internet. The work involved in creating and teaching such carefully structured courses does, however, create a significant workload for educators—a fact that many universities do not seem to want to recognize by reducing the required hours for e-teaching. Such courses are frequently run by one educator for each group of students. Low technology-constructivist online courses are usually not offered for distance education on a large scale, but for quality education at universities and companies (Weller 2002, 145–53).

American Cultural Studies Onweb (ASC-onweb.de): background

In the context of fundamental changes in the field of American cultural studies as well as the realm of information technologies, *American Cultural Studies-onweb* (ACS-onweb.de) defines new content and methods in teaching in a national and international framework. The ACS-onweb project grew out of an initiative in the German state of Baden-Württemberg to further innovative teaching based on the necessary intensification and internationalization of tertiary education.

Major educational policy in Germany is formed at the state, not local level, and throughout the German educational system, a fundamental restructuring of all disciplines as taught at universities is taking place, with the state of Baden-Württemberg often leading the way. Currently, especially due to the influence of a move to standardize educational systems within the European Union, a bachelor's degree (BA) has been introduced to the German university system in many disciplines. Until recently it has had no undergraduate degree such as the BA, with the Magister Artium (MA) or Staatsexamen (tertiary degree for teachers at secondary schools) as the first degree granted after a minimum of eight semesters at university. One aspect of changes in policy, curriculum, and degrees awarded has been a significant reevaluation of English as a subject in general and specifically the introduction of cultural studies. It is an innovation that cultural studies now occupy an equal plane with the study of literature and linguistics. In addition to a changed focus in content, the intermeshing of knowledge, methods, and educational strategies as well as the sustainable presentation of research and pedagogical instruments are at the core of this changed perspective.

American cultural studies have undergone constant change both in the United States and Germany. The most significant change has been the move away from a canon of texts and topics, with an attendant move toward a canon of methods that supports the learning process within a larger context. Rote knowledge has become less important; rather, students must learn how to access and process information for learning in the present and in their careers in the future. As a team of a German (Holtkamp) and an American (Fischer-Hornung) with many years of experience in teaching and researching American (cultural) studies within the context of German universities, we concluded that our students should be introduced to the complex intermeshing within the field of cultural studies as well as that of the Internet. With its own unending network of information, the WWW provided an ideal tool. It would allow us to introduce students to the information available about the USA on the Internet, enable them to learn how to utilize the Internet for scholarly purposes, and also make the Internet itself a subject of their scholarly endeavours.

For students in the field of cultural studies, it is important that they understand the symbolic structures inherent in social contexts in order to understand the cultural context of the United States. This holds particularly true for students outside the United States since language, particular semiotic systems, and the associated communication media are increasingly significant not only within the USA but also globally. As human beings, we are called upon to process and evaluate information with various characteristics and forms transmitted by various media in the most diverse situations. For American cultural studies it seemed reasonable to make the complexity of socially grounded forms of information about the USA the object of investigation and to develop suitable didactic methods to transport this content as an integral part of our teaching. In a communication society this means using information technology in our teaching.

The courses developed in Heidelberg and Stuttgart—a three-year project supported by the Ministry of Science, Research, and the Arts of the State of Baden-Württemberg as well as the Universities of Heidelberg and Stuttgart—are part of the process to initiate changes in the curriculum and to fill these new requirements with content and to develop appropriate methodologies as well. The grant provided by the ministry and matching funds provided by our respective universities enabled a one-and-a-half-year development and implementation phase; thereafter a phase of comparable length was devoted to refinement of the original courses in a second run as well as to the development of comparable new courses. During the initial phase, our positions were fully dedicated to developing the ACS-onweb project and we were provided with two graduate assistants for content development (Thorsten Gutmann, Stuttgart, and Susanne Porr, Heidelberg) as well as two graduate assistants for software development and management (Jörg Bäuerle, Stuttgart, and Fabian Lorenzen, Heidelberg). In the second phase we retained the help of our graduate assistants, but returned to our initial teaching positions, thereby testing the practicability of e-teaching within the context of our regular teaching load.

Already during the first phase of the project, Stuttgart developed its own open-source e-learning platform (summer semester 2002) based on PostNuke and subsequently Heidelberg also adopted an open-source platform, MIT's DotLRN (summer semester 2003).[1] The project was structured from the outset so that the latter phase would entail a second run of the courses we had developed, the development of new courses, as well as the expansion of the international dimensions of the project.

The increasingly rapid internationalization of the academic community moved us to internationalize our programme, and we contacted various universities abroad. Initially, the availability of WebCT as a password-protected e-learning platform at the University of Heidelberg enabled students from various universities to meet in our virtual classrooms. In addition to the course developers in Stuttgart and Heidelberg, each partner university had a coordinator assigned to the project: Ingrid Day, Department of Communication, Information, and New Media at the University of South Australia, Adelaide; William Boelhower, Department of Germanic Languages, American Studies at the University of Padua, Italy; and Irina Shemelyova, Faculty of Philology at the State University of St Petersburg, Russia. The transfer of credit within the systems is recognized by our partner universities and is formalized either by agreements within existing partnerships among our universities or within the ERASMUS program of the European Union. Negotiations have been initiated with several North American universities in order to take American cultural studies—after its diversion and possible reprocessing from various perspectives—back onto its "home turf," so to speak.

The various international perspectives on US culture yielded new approaches to and perspectives on the study of the US sociocultural context implicit in the American cultural studies approach. In addition, students from very different cultural backgrounds learned to formulate, present, and discuss their arguments in the virtual classroom. The participating students in the humanities attained specific knowledge about US cultural studies and simultaneously acquired Internet competence, skills that we assume will be beneficial to them in their future careers. With these goals in mind, we have embarked on the project of creating an expanding global network of online American cultural studies.

Already in the test phase, the e-learning courses were offered as equivalent to the standard curriculum and credits were allocated according to the system at the respective partner university: currently the courses are credited for a full term, based on an intensive course of four hours for the seven-week duration. Ultimately, it was not only a project developed by university instructors, but also by our students. Students knew, since they were a part of a pilot project, that their input could directly influence both the form and content of future courses. The courses were continually optimized, based on feedback from students and our partners at the participating foreign universities. ACS-onweb proved to be a flexible, adaptable core design for courses that could be effectively used to teach an expanding number of topics in American cultural studies and other areas in the humanities as well. It also provided a basis for working with our international partners within virtual cooperatives.

Conceptualization and design

The courses are designed to reflect a multiplicity of topics and methodologies in approaching the field of American cultural studies. Students who often have only rudimentary knowledge of the United States (often based on stereotypical assumptions), who have little experience in interdisciplinary approaches, and who may have little knowledge of information technology are given the opportunity to expand their knowledge and develop these skills in e-technologies and interdisciplinary methods. The courses were structured to support individuality, team-competence, goal-directedness, openness, cooperation, coping skills in a demanding learning environment, and transferability of knowledge and skills to other tasks. With these ends in mind, the following seminar modules were developed within the initial 18-month phase of the ACS-onweb project:

- American Culture and Identity (Heidelberg)
- Ethnicity, Race, and Immigration (Heidelberg)
- Approaches to American Regionalism (Stuttgart)
- The Situation of the American City Today (Stuttgart)
- The Theming of American Culture (Stuttgart)
- Youth and Media Culture (Heidelberg)
- Sports and American Culture (Stuttgart)
- Gender in US Society (Heidelberg)

In the first semester of the project the instructors were introduced to the use of e-learning platforms in a course offered at the University of Heidelberg and simultaneously texts, links, and visual material were compiled. In order to involve students from the onset, a class called American Cultural Studies Onweb was held at Heidelberg in a computer lab, during the course of which students helped to research and discuss available Internet material. They defined their interests and developed small lessons on topics such as race and ethnicity, youth and media, and gender.[2]

Based on the experience in this first test and development phase, which indicated that intensive immersion in both content and technology would be the optimal approach, we designed a seven-week crash-course model. The first week consists of an orientation and dry run to familiarize the students with the various technical aspects of the course. The students receive a password to enable access to the e-learning platform. During orientation, an insystem tutorial helps the students to access the syllabus, test all the functions of the platform, access their course material and the class calendar, utilize the internal e-communication features (forum, e-mail, and chat), access and submit assignments as well as material they would like to share with the class (for example, files containing texts, hyperlinks, visuals, audio-material). For the instructor designing the course, all these features are available as modules in the e-learning platform (for example, communication tools or calendar) or can easily be uploaded into the course as needed by choosing from the palette of open-source platform features.

We learned that it takes time to get used to the new learning environment since students not only have to tackle new content, but must also meet the challenge of encountering a totally new media based teaching methodology. Since e-learning platforms are increasingly user-friendly and our students generally use the basic feature of e-communication in their daily lives (e-mail, instant messaging, and the WWW), the introductory "dry-run" should become increasingly easy in the future as the Internet becomes an everyday feature of our lives. After the introductory "dry run" during Orientation (week one), the remaining six weeks of each class are divided up into three course sections consisting of three major topics that are in turn divided into subsections. For example, the course The Theming of American Culture, is divided into the units Museums, Malls, and Themed Environments, and the course Youth and Media Culture has the submodules Advertising and Youth Markets, The Internet and the Media, and Youth Groups and Hanging Out Virtually.

The course material in Stuttgart consisted of texts that were scanned into the system, whereas the course content in Heidelberg was entirely Internet-based, consisting of material linked directly from the WWW. The texts or Internet links for each of the three sections are made available to the students in a hierarchically structured web site; they are guided through the material using the assignment feature. They submit their answers virtually (four to five pages of text for each assignment) as a file within the e-learning platform. These assignments are corrected and graded by the course instructor and are handed back electronically within the student file feature of the course. Part of every assignment is an independent Internet search task to develop skills in scholarly Internet research. The students do their background research using the Internet, evaluate what they find, determine what they consider the qualitatively best links, and make this information available to their fellow students. Simultaneously, the topics and associated questions are introduced via the assignments and are discussed intensively in the Forum as well. This link between guided research assignments and simultaneous discussion as a group in a virtual seminar forum enables the participants intellectually to link what they have learned in their Internet search with the work of their fellow students in the international classroom.

In addition to work on assigned material and active participation in seminar discussion in the Forum, students form (by signing up in teams according to the interests they develop in the first half of the course) or are formed (by random selection) into internationally mixed teams. Their team cooperation in producing an Internet presentation in a virtual international team is part of the final phase of the seminar. The Internet presentations they submit are then posted on our open ACS-onweb project web site. Our project's homepage has links to presentations at each partner university. The knowledge that this contribution is their "claim to fame" on the Internet as well as the recognition that they can potentially add this public presentation to their CV is, of course, a highly motivating aspect of the ACS-onweb course design. The open availability of course results also potentially contributes to the long-range accessibility and sustainability of the ACS-onweb project, making material available to other students and instructors throughout the world, creating a further level of international networking.

ACS-onweb enables, in contrast to the fixed structure of traditional classes, an interactive, flexible form of taking and teaching classes that is independent of time and place. This has been noted as one of the prime advantages of an online class and several students with special needs (for example, parenthood, work that demands travel, disabilities) have signed up to take a second seminar. In addition, the students know it is essential that they learn to use the new media, proficiency in which will surely be an advantage in their future careers. The information they access is up-to-date, with almost unlimited sources of information from texts to visuals—a range of material no single textbook can offer. The American Memory Project from the Library of Congress is one outstanding example of invaluable archival material available online.

As indicated in our discussion of the constructivist approach earlier, ACS-onweb has specific characteristics that fit into modern concepts of pedagogy. The course's emphases on student/instructor and student/student interaction as well as its focus on process and interpretation are factors in the student's own experience and understanding. After their initial introduction to working with materials on the Internet, students can find their own material and present it to their fellow students for discussion. In this way the course content extends well beyond the initial input based on a selection of texts and links thought suitable by the instructor, expanding to include material submitted by the students themselves. The e-learning platform enables direct and immediate contact amongst the international participants in the virtual classroom and enables an increased sensitivity to intercultural communication strategies and styles. The e-technologies (including the Forum, e-mail, and chat, support dialogue) and the work on an Internet presentation in a virtual team support collaboration—as a result, learning becomes a personal as well as a social activity.

The constructivist approach casts the educator in the role of facilitator and mentor. The feed-back students get in these e-learning seminars comes from the instructor in reference to their written submissions and also from their fellow students as well as the instructor in the Forum or via e-mail. The instructor not only facilitates individual learning but also actively participates in class discussions, enabling him or her to facilitate the content and form of the group discussions. Finally, the flexible structure of online courses allows students to have ready access to the course; this flexibility in turn affects course content and form. It enables a constant interchange between the latest developments in the field and the didactic transference of research into course content. Grades are given with equal weight to the following activities: written assignments, contributions to the Forum, Internet search task, and team Internet presentation.

Problems and potentials

Internet accessibility available to the students in the various countries differs greatly. The WWW may be worldwide, but access differs radically based on economic differences among user nations. To enable access for our students in St Petersburg, for example, a fund was set up to pay for access through an Internet café near the English department. We could not assume that they would have access to the Internet in a university PC lab or have private access. In addition, the time zone differences between Germany, Russia, Australia, and the United States were not always easy to negotiate, but a core time where everyone could expect to be able to participate in chat sessions via Instant Messaging was established.

The great advantage of working online, of course, is flexibility, with access at the university, in an Internet café or at home. After the initial dry run period during the first week of class, students can and did log in to our virtual classroom at all hours of the day and night. Soon students tended to access the course seven days a week. The size of an e-seminar proved to have a significant effect on the success of a class. Since input in the Forum, for example, tends to increase in frequency during the course, the class can become quite unwieldy with increasing numbers of participants. In one class of seventeen participants we had 463 postings, some of them quite lengthy, within six weeks. With 20–25 participants the Forum became increasing unmanageable. A class size of approximately fifteen proved to be ideal; a fact that has been substantiated by colleagues at various conferences. We suspect that the vision of some university administrators included an increase in class size in e-learning seminars; just the opposite has proved to be the case if quality education was to be achieved in a seminar format.

The possibility of integrating international students in the virtual classroom enabled a differentiated level of discussion of US culture and society. We have had US participants from the onset since many are exchange students at our respective institutions in Germany. They have frequently availed themselves of this opportunity to learn about their own society from the "outside"—both in the different perspectives offered by the instructors as well as new insights provide by students of diverse nationalities. In addition, we have been able to offer courses crossing the borders of disciplines and departments. Students in Heidelberg and Stuttgart can, for example, now enrol in the Department of Communication, Information, and New Media in Adelaide, and Adelaide students can participate in our American cultural studies seminars.

We were also able to arrange "expert chats," making experts in the field available directly to the students—one example was the chat with Teri J. Edelstein, museum consultant in strategic planning and program development and former Deputy Director of The Art Institute of Chicago, during The Theming of American Culture seminar. In the context of increasing globalization, such internationalization of participants and resources utilizing e-technologies seems particularly relevant. Other educational approaches can also easily be integrated into our approach. For example, in order to validate and concretize the relationship between virtual and real space, Wolfgang Holtkamp organized an excursion American Realities/American Fantasies for the best participants from both Stuttgart and Heidelberg. Themed environments became not only a virtual but also a material reality and, of course, greatly enriched the learning experience for the students. He also tested a mixture of face-to-face and online teaching at the University of South Australia in Adelaide. This mixed structure enabled immediate feedback from the Australian students, who put particular emphasis on the positive effect of having an international classroom situation via the Internet. A project to bring together Australian and German students who have participated in one of the ACS-onweb classes has been planned by Ingrid Day and Holtkamp.

There have been several spin-offs from the project, such as Dorothea Fischer-Hornung's integration into the e-learning team at the University of Heidelberg and also her work as a member of the advisory board for the development of MIT's open source platform DotLRN. Wolfgang Holtkamp, in turn, has developed an essay-writing course called English Composition Online (ECHO) to be taught in Stuttgart.

Conclusion

Based on the positive experience in the development phase of ACS-onweb, online courses in American cultural studies have become an established part of the curriculum of the Universities of Heidelberg and Stuttgart, as well as our associated partner universities. The solid integration of these courses recognizes the fact that the age of information technology demands the exploration of new content and the internationalizing of form made possible by e-media. With careful planning, design, and "grooming," the Internet has proved to be extremely useful for teaching in the humanities, both as a source of content and as a basis for didactic methodology. The course content and teaching methodology for eight courses have been established. These courses will continue to be offered cyclically in continually updated form and additional courses will be developed.

For our students, the project has provided a means to learn about the latest developments in American cultural studies as well as to develop their Internet expertise. Concrete contacts, friendships, and a virtual exchange programme have been established—turning the virtual classroom into a real-world resource. Therefore, we will continue to expand our international cooperation across departments and disciplines, because we see this as the real potential of the Internet, forming a virtual centre of international American cultural studies. For us as teachers it has been an intellectually and personally enriching real and virtual adventure to work in a totally new teaching environment.

Appendix

Two studies of our course have been conducted within the last year and an additional virtual English project is being launched. We have included them here as an appendix to our paper.

We were consulted by Simon Mangler, Matthias Mechler, Benjamin Reichert, Arne Spieker, and Florian Wildemann, students at the Wirtschaftsoberschule in Baden-Baden (secondary magnet school in economics). This team interviewed Fischer-Hornung and then developed a questionnaire sent out to participants in former classes. Based on the twenty-six forms which were returned (about 50% return rate), this is a summary of their results, submitted as a final school project entitled, E-learning der Universität Heidelberg. Their project essentially confirmed our own debriefing of our students. Since neither their questionnaire nor our own face-to-face and written feedback have any claim to scientific validity, we have included a summary of their results as an appendix.

Students said they accessed ACS-onweb an average of five times a week and 95% at home. 78% felt they improved their PC/Internet skills. The same percentage thought that the flexibility offered was also very important; the remainder thought it was advantageous. 82.6% felt that an online course makes it easier for them to complete their studies. 65% saw improvement in their communication with their instructor, with 30% reporting it was about the same quality as in a face-to-face seminar. However, only 39% felt it had a positive effect on the communication with their fellow students; almost 22% felt a negative effect on student interaction: "there's nothing like talking to someone in real life and have a cup of coffee after the seminar ☺" (quoted in Mangler et al. 2003). Although our students stated that there is nothing like having a "cup of coffee" together after class, 65% did feel they had better contact to students based on our virtual international partnerships, where direct contact is not possible. 61% felt their soft skills (such as teamwork, independence, and creativity) improved. 74% intended to take another ACS course. The only area where they felt there was no significant improvement was in their language skills, with 8.7% reporting significant improvement, 47.8% reporting some improvement and 43.5% reporting no improvement. This result cannot be surprising, since the ACS courses were not designed to improve language skills, but rather to teach American cultural studies. Any improvement in the students' English is indirect and due to the fact that the course is conducted in English and the assignments are returned with linguistic feedback in addition to content corrections. The workload amazed a good number of students and this is reflected in a drop-out rate of about 25% overall (not more than in a usual face-to-face class in our experience).

One of the greatest disadvantages to online teaching seems to be the lack of body language signals, reflected in the following comment by one of our Australian participants: "I learnt that other cultures' command of English is quite good; however, they misjudge the nuances given to words or sentences. Even with skills employed through our use of communication through the Internet, this becomes difficult as diplomacy is helped by the use of body language and our other senses, which is why face-to-face communication becomes meaningful" (quoted in Mangler 2003). We find these observations held true for our participants in general, and to offset this effect we made a slate of emoticons available for inclusion in e-mails and postings.

Another direct result of the development of tandem language teaching is being developed by Sabine Gläsmann, as part of her Ph.D. thesis for the University of Sheffield. She will involve students of German at the University of Illinois, Urbana-Champaign, and students from the English department at the University of Heidelberg in a virtual language exchange. In a guided environment, they will exchange language-learning skills and projects in real time based on written e-communication as well as aural and oral communication via web cams and headsets. The project will be divided into three phases over a period of nine months.

Last and very certainly not least, we would like to point to the most extensive study of ACS-onweb, which is part of a doctoral thesis to be submitted at the University of Heidelberg by Nicole Flindt, who provides an extensive comparison and analysis of the courses in Stuttgart and Heidelberg, as well as twenty-two other projects. In correspondence with Fischer-Hornung, Flindt makes the following comment on her evaluation of ACS-onweb: "I evaluated 23 e-learning courses in my thesis. One of them is the ACS-onweb course Race, Ethnicity and Immigration (20.10.–20.11.02). Because the concept of the ACS-course is so different from the 22 other courses I could not evaluate your course with my normal criteria. I preferred to give a written comment and hope you like to hear that I will say that the didactical concept as well as the work in international groups are awarded with the [highest German] grade *sehr gut*" (Nicole Flindt, e-mail message to Dorothea Fischer-Hornung, 27 September 2003).

Notes

1. In comparison to proprietary e-learning platforms (such as WebCT and Blackboard), Postnuke and DotLRN (http://dotlrn.org/) offer a "low budget" open source alternative to commercial products without sacrificing the complete slate of features required for the ACS project. The features of these open source platforms, moreover, can be expanded or reduced as required.
2. The results of this first course can be accessed at: http://www.rzuser.uni-heidelberg.de/~el6/. The content the students generated was factored into the first online seminars taught in subsequent semesters.

9
Transatlantic Exchanges: Mediating Student Learning through e-Discussions

Duco van Oostrum

And then there was geography.
(DeLillo 1972, 30–1)

In 1998, Judith Hakola (from the University of Maine, Orono) and I thought we would add an extra, virtual dimension to our respective American sports literature and film modules—we would have the American and British students communicate with each other. Neither of us is by any stretch of the imagination an IT expert, but we figured a bit of e-mailing should be within anybody's reach. In 2004, we're still at it, completely engrossed by all the technological wizardry of WebCT, video conferencing, and other modes of virtual teaching. The idea of the transatlantic bulletin board, however, remains virtually unchanged; communication between students from different cultural backgrounds, focused around similar teaching material, enriches the student learning experience far beyond the normal seminar setting. This article presents the specifics of using a bulletin board to teach American sport literature and film. Nevertheless, we would expect that the underlying principles of our course would transfer to different disciplines and cross over into other area studies as well.

The relative simplicity of a bulletin board naturally attracted us. The implications and ramifications of using this virtual reality proved to be anything but simple. This essay discusses the experiences and outcomes teaching a transatlantic exchange course in different sections, using illustrations from the bulletin board. Along the way it considers crosscultural IT challenges; the experience of teaching within different education systems; aspects of American studies; and

Don DeLillo's novel *End Zone* (1972). My concluding remarks emphasize some of the possibilities and challenges bulletin board activity present in an English literature setting in the UK.

Transatlantic IT demands

Before we even started, we were quickly testing the limits of teaching with IT. At both institutions, there were different communication programs, and it was also quickly apparent that US users were more familiar with e-mail. The first years were done on an open University of Sheffield web site (appropriately named End Zone: http://www.shef.ac.uk/english/modules/lit356/site/ezdisc.HTML), where a live link would connect to a bulletin board for posting messages and responses. At the University of Maine, students used the course management system FirstClass for internal communication, but at Sheffield there was no such system in place. It was also not possible to configure FirstClass to grant open access to it. The greatest challenge of all, we quickly found out, was distributing usernames and passwords and instructing the students how to use the facilities. The students at the University of Maine had new postings appear in their regular e-mail inbox after the initial set up, but Sheffield students always had to log in to the bulletin board itself. Within WebCT, the Maine and Sheffield students have to log in to a dedicated WebCT course set up for both classes. While we could employ commercial bulletin board providers, we preferred to remain within institutional boundaries. The current version of the bulletin board functions magnificently, and only rarely does a server or connection let us down. We do need special permission from WebCT, however, since licences are institution based (we were set a limit of 100 participants in 2004.)

Experiencing different educational systems

The bulletin board allows students from either side of the Atlantic to experience different types of university education. Not the least important is the different academic year. Trying to coordinate the respective classes remains one of the biggest challenges in this project. The modules both run in spring semester but the dates vary considerably. Sheffield starts after the Maine semester is already three weeks underway. In the spring, Maine goes on Spring Break for a week in March and Sheffield has an extended three-week Easter holiday. Maine then ends its semester when Sheffield still has four weeks to go in its module. In a twelve-week teaching schedule, our institutional schedules only allow for about five weeks of intensive contact. On the actual bulletin board, students regularly comment on the different academic calendars; Sheffield students become knowledgeable about the iconicity of "Spring Break" while Maine students reflect on huge gaps in the British semester schedule.

The four-year programme of the US higher education system, with requirements across disciplines during the first two years, contrasts markedly with the discipline-specific focus of the three-year UK programme. In our case, Judith Hakola's course, English 249, can attract up to 100 students. As a second-year course, the course caters to students from all disciplines who use this module to fulfil part of their general humanities requirements. The course has three contact hours, taught on two days to the full group. In fact, the English majors and minors are typically in the minority. The module at Sheffield, English literature 356, is exclusively for English literature and American Studies students (or "dual honours" students). As a final-year "approved" module, it caps at twenty-four students, is taught in two groups of twelve, and has two contact hours. The assessments for both modules reflect this academic difference as well, with book reviews, quizzes, and an analysis making up the bulk of assessment for the University of Maine module. At Sheffield, the module is assessed via exam, extended research paper (3,000 words), a project, and short assignments. In both cases, bulletin board activity "counts" in terms of participation, but I have recently increased and formalized this aspect. In short biographies and introductions on the bulletin board, the various differences between the UK and US students were frequent topics, and these discussions of difference proved an interesting unexpected result of the bulletin exchange. In a microcosmic manner, the bulletin board allowed a virtual experience comparable to that of a formal physical student exchange.

The subject area: education and sport during the US and UK exchange

While both our modules are taught within the academic discipline of English, the subject area, American sport literature and film, is not at the core of the English literature curriculum. Even having the module accepted as part of the curriculum requires stringent validation procedures. From the margins of the discipline, it is perhaps easier to venture into other disciplines and teaching experiences, and we consider this topic ideal for student exchanges. Sport itself is an under-researched genre in the humanities, and teaching it as an academic discipline with limited reference material poses a particular problem. The link between sport and nation makes a module of "American sport" challenging to nonAmericans. Many students, for example, comment on the insular aspect of national sports: English students are often unfamiliar with rules of American football, baseball, and basketball, while questions about cricket and rugby emanate from the other side of the ocean. The "American-ness" of US sports literature and film features significantly in most texts (as in much American literature), and in many cases there are overlaps between American cultural mythologies and the experience of sport itself.

In its ability to overcome cultural divides, the bulletin board proved itself as an educational tool. The transatlantic exchange bridged the unfamiliarity with American sport experience (in spite of its proliferation in American popular culture) for the UK students with the lived experience of the US students. It also served as an opportunity for the US students to reflect on their experiences, making the familiar suddenly unfamiliar. The additional focus on literary analysis seemed to make texts literally come to virtual life. Many of the students taking the module at Maine were themselves student-athletes, a particular object of investigation in the Sheffield module. This link between sport and education in the US is unfamiliar to the UK students. Through the bulletin board, they could compare experiences of sport and education. The topic of "the student-athlete" was the most popular and contested of the bulletin board. This topic also relates to the central text of the module but the discussions frequently ventured into personal academic and sporting experiences. Two examples (names withheld and text unedited):

1. A helpful U Maine student explains:

> the university of miami football team draws on the average about 60,000 fans per contest and has a budget of over 20 million dollars, or 15 million pounds. that is the football team alone. the university

of maine's entire budget consists of only 15 million dollars. so some schools are definitely bigger than others, and most of that has to do with the population of the area the school resides, there are more people in the city of miami than in the entire state of maine. as far as the cold war goes, well, it doesnt matter. football can be a very violent sport. and sometimes, when a player has no escape when football is not in his life, violence can be that escape. now, this doesnt mean that football players are all violent people, but the stereotype would tell you that. the stereotype would also tell you they are dumb and slobs and so forth, that is not true. i would say the percentage of idiots to smart guys on a football team equals that of the percentage of idiots to smart guys in real life. this book really brings out the stereotype of a typical football player. especially one who has failed to play at the big-time schools because of his inability to grasp real life away from football. it happens in every sport. well, i hope this answered some of your questions. ill be here at my house all of break, so you can write me. as for now, bye . . .

2. A Sheffield student writes about sporting life at a UK institution of higher education:

There was a question about how sports operate in Britain within the university system which i will endeavour to answer. Basically it isn't taken half as seriously! There is no money at all in university sports. You come to uni and then if you fancy playing soccer or rugby you join the relevant club. There are very, very few schools that offer any kind of sports scholarship and people join the clubs often just for the social aspect. If you can play to a reasonably standard then you will be selected to play in the 1st or 2nd team but even if you are hopeless there is usually a 4th and 5th side depending on the size of the club. There are no employed coaches and the players elect a committee to run the club. There is a club for pretty much any sport you can think of from soccer to skydiving and you pay a nominal fee to join and then a small fee every time you play for the hiring of buses etc. The universities play each other in leagues dependent on where you are in the country and there are cup and plate competitions. If the team is very good and get through to a final you may get as many as 50 spectators {!!??}—not quite 50,000.

> There is no special arrangements made for those on teams—you basically either don't play or miss a few lectures! If you want to play professionally then you join an outside club and hope to get scouted although there are some ties between universities and local clubs. I hope that this explanation was of some use—it is certainly very different to America!

I have quoted these two entries at length because they in many respects typify the exchanges. First, they illustrate the difference in tone and language in messaging between the UK and US students. The US students used the bulletin board as part of their daily messaging, ignoring capitalization for example. The UK students were much more formal on the bulletin board. The relevant cultural information students receive on the subject area (American sport literature and film) via the bulletin board is magnificent. The number of entries trying to dispel the notion of "dumb jock," for example, is impressive. The Sheffield student finds her/himself exchanging ideas with the topic of discussion—"the" American student-athlete. The American student-athlete, in turn, commonly identifies himself as such: for example, one student describes himself by the position he plays on the football team and his number: [Name], Fifth year senior, Linebacker #58, major (teaching and coaching).

The text: *End Zone*

DeLillo's text forms the central focus of the discussion—the reference in the bulletin board entry above refers to "the book". The hilarious first entry from one of my Sheffield student was:

> Hi, I would like to ask what exactly what the term "endzone" is in football. We've just discussed the DeLillo book in our tutorial, and a lot of the language he used went straight over my head. I'd also like to know how students are selected for university—is a talent for sport really of major importance at Maine? I hope you can sort me out!

The student indicates immediately the relevance of football terminology to reading DeLillo's book. With specific American football terminology dropping in on most pages, it is difficult to teach

End Zone in the UK without fumbling. In most of DeLillo's work there is immense play with signification, with empty referents, and with conspiracy theories in a postmodern world. The football field, with its enclosures, its rules, its yards and the idea of progression and forward motion, becomes part of an enclosed signification process on and off the actual playing space. As Tom LeClair (1988) has argued, DeLillo's books frequently revolve around the limits and possibilities of closed systems. On the Texas plains, football is religion, and DeLillo not only plays with the "American-ness" of the setting but also with stock cultural characters (the Jewish football player, the Public Relations man, the black jock, etc.). These elements both ground the novel in American experience and raise the issues of translatability. Will UK students see the athlete's identity as something foreign and other? Will US students recognize the type as a viable identity even off the field? Does a different reading of the sports metaphors in the text also lead to different readings of themes within the novel? How can one productively read the text from both angles? Can this text be read "correctly" if one doesn't know American football and sports?

These were some of the ambitious questions we set ourselves to answer in our respective sports literature classes. It turned out that DeLillo's *End Zone* was a suitable text to confront issues of cultural location and reading because the novel itself appears to split into a world of "real" college sports at a small West Texas college and a surreal postmodern world of Baudrillardian "desertification," where all meaningful signification has ended in languages of the "untellable" of postnuclear fallout. It is possible to read the text as the story of a student-athlete obsessed by football, who has to find his place in the team and the university. It is also possible to read the novel as a description of language games in which language itself is the subject because it describes and conceptualizes an "untellable" future. In other words, sport and something that sport is supposed to represent are in direct conflict with one another. The obsession with the "untellable" of violence in language and the world in DeLillo's novel was highlighted in the bulletin board discussion through different kinds of readings. For the Sheffield students, part two of the book, the "play-by-play" commentary of a match, proved indeed "untellable"; we even had to discuss "who won." By contrast, the theoretical game play of *End Zone* (similar to Beckett's *End Game*) was appreciated for all its postmodern signification. The Maine students read the novel especially for its depiction of the real world of small college sports. There was an insistence on a realistic description of sport that was also apparent in another text, John Edgar Wideman's "Doc's Story" (1990). As a student from Maine put it:

> I would have to say that it is possible for a blind individual to play basketball but not at a competitive level. Like many sports, basketball requires hand eye coordination, which unfortunately, not everyone is able to do. It seems harsh to discriminate in this sort of manner, but in real terms it is simply unrealistic to make the assumption that blind people can play the game of basketball as we know it.

The insistence on the real game contrasted with the attention on narrative and fiction of the Sheffield students. On the bulletin board, these two ways of reading sometimes clashed, but most often, the students discovered ways of reading that would have been difficult to obtain without having had the transatlantic experience.

Conclusion: the transatlantic bulletin board and teaching English

An English Subject Centre report on the use of IT in English departments in the UK notes that bulletin boards were a frequent topic of discussion among lecturers interested in pursuing IT in their classes:

> During a discussion of small group or seminar teaching, the advantages and disadvantages of electronic discussion boards were frequently a common topic of debate. The concerns raised against their use included:
>
> - they don't allow lecturers to assess adequately student participation;
> - they promote plagiarism (students use other student ideas posted during discussion);
> - they result in disruptions (inappropriate language; digressions).
>
> While raising these objections, the opponents to electronic discussion boards did not consider these matters as concerns

and issues commonly experienced in seminar meetings without discussion boards:

- It's difficult to assess student participation (do lecturers assess seminar participation by the simple fact that students attend class? Or by how much students contribute to discussion?).
- Plagiarism is always already a concern (one mundane example, the regurgitation of discussion in written assignments).
- Seminars require strategies for controlling disruptions (students arriving late; individuals dominating discussion; or unwelcome or unwanted distractions disrupting discussion). (Hanrahan 2002, 26–7)

Participants recognized that the use of an electronic communication tool involved the engagement of issues and problems that are sometimes simply taken for granted. While the above is a single example, it points to the ways in which the use of technology provides an opportunity to reflect on teaching and learning.

In this ongoing case study, we have enabled an enriched learning experience for students of American literature that transcends the experiences of the isolated classroom bulletin board. In the transatlantic version, students add to the classroom experience in ways that do not compete with seminar learning, as some of the opponents of bulletin boards seem to suggest. The students in Sheffield usually revise prejudices about student-athletes in the US and are more informed about different ways of reading literature according to context. The bulletin board creates an informal learning environment of genuine exchange between students, rather than a teacher-led exercise in interpretation. It also allows for good fun.

The bulletin board is an under-used IT facility in English studies. In addition to the academic value to learning as outlined above, the bulletin board helps satisfy the incessant demand of measurable transferable skills in English literature teaching in the UK. The interactive element of the facility makes it an excellent tool to measure student progress. The bulletin board provides a record of a type of self-reflexive student-learning that can be submitted as part of the formal assessment that can be monitored by external examiners if need be. We have now formatted the bulletin board in WebCT in such a way that contribution can only be made in weekly topics, with other topics "locked" until the syllabus progresses to that topic. As a result the bulletin board exists alongside WWW resources.

The initial logistics of the bulletin board remain daunting; the closed VLE of WebCT limits the open-ended play of normal websites and there is a demand for a computer programme that allows for this type of exchange within the VLE without all the permission procedures. Once the bulletin board is in place, however, the rewards are both measurable (the content and discussions) and immeasurable (the enthusiasm for the subject and liveliness of seminar discussion).

Part 4

Hypermedia: Theory and Practice

10

Audience, Purpose, and Medium: How Digital Media Extend Humanities Education

Eric S. Rabkin

Not so much credo as observo

The training of students to evaluate resources and compose in new media forces us all to confront the practical pedagogic and theoretical aesthetic issues behind the uses of those media. This has felt true to me from the time I first used electronic discussion boards to augment my lecture classes in literature in 1975 until today when I regularly teach two University of Michigan computer lab courses enrolling undergraduate and graduate students explicitly interested in new media. In English 415/516 Technology and the humanities (usually team-taught with Victor Rosenberg from our School of Information) and English 414 Multimedia Explorations in the humanities, the confrontation of diverse media makes issues vivid that have in more traditional courses too often fallen below consciousness for many people.

In every humanities course, so long as it requires of students at least the production of an essay, the course should to some extent be teaching literacy. It seems to me that whenever we teach literacy, no matter what else we are doing, we are also usually trying to show people how crucial it is to develop—and work within—a firm sense of audience and purpose. That is of course still true in teaching people to use, evaluate, and create digital productions in digital media, but digital media, both because they feel new to us and because their capabilities keep developing, problematize the focus on audience and purpose, making us aware of medium itself, and thus potentially enrich enormously the conscious educational enterprise. For instance, we now must deal with the fact that choice of medium influences audience expectation. Few instructors expect paper essays to include graphics, although that would be easy enough to do today, yet all viewers of web pages do expect graphics. What happens when you do or don't meet audience expectations in any given instance? How free is one's choice of medium? To what extent do such choices flow from a sense of audience-and-purpose? To what extent is site design an act of setting media expectations, and hence crucial to the argument inherent in a given web site? Just as there are different sorts of books (for example, reference books to be dipped into, novels to be read sequentially), there are different sorts of digital productions (for example, web sites to be dipped into, web sites to be read sequentially). However, digital media offer new possibilities (for example, web sites that contain up-to-the-instant content, such as the current time or the current age of an author being discussed; interactivity, such as requesting and then employing the user's name in the web page text, thus allowing for a dramatic form of direct address; random variations, such as pictures of one's cat at different ages appearing every few seconds in text discussing the metaphor of "nine lives"; and so on). The set of expectations an audience has about any of these new media is necessarily more contingent than those raised by book culture. And yet these digital media build conventions about themselves, just as books have. The current attacks on PowerPoint (Tufte 2003) complain not about PowerPoint as it might be used but only as it usually is. Humanities education should allow not only the critique and creation of works like those we find but unlike—and sometimes better—than those we find.

Humanities education must extend itself beyond sequential literacy to deal with more capacious media and with diverse and flexible expectations for production and consumption. Perhaps most central to the evolution of expectations within this new information environment is fully accepting the notion that intellectual products are the result of the work of many people—including often the reader, now perhaps better called the user—over an extendable time rather than the work of one or sometimes two or three accomplished over a fixed time. That is, digital media, functioning as they do in the world of networked computing, often break down the boundaries we once took for granted in setting tasks for our students: the finality of composition, the identity of the author, the role of the audience, and the unity of purpose. We live and work in an inherently collaborative infosphere. As always, it is the task of the humanities to make the world we inhabit as rich and good as possible. Today that means bringing digital tools into our traditional workshop.

The humanities tradition carries on

For many, "the humanities" is synonymous with the concerns of university departments organized around the practice of criticism. Matthew Arnold famously defined criticism as "a disinterested endeavour to learn and propagate the best that is known and thought in the world" (1953, 162). Arnold's vision is fundamentally conservative. It argues explicitly for selection among those items that have been known and thought and implicitly for two sorts of education: the education of the skills necessary to distinguish the best from the rest and the development of those skills necessary to make one's grounds for distinction clear to others so that those best items will enrich the widest possible audience and thus spread in our culture, if not across classes then at least through time. The critic is in many ways like an athlete, or an ancient Greek *rhetor,* ever practicing a known art so as to become ever more its master and through that mastery perform for the public. Perhaps the most glaring exclusion here is invention. Arnoldian criticism does not seek to discover something fundamentally new, as science does, but to learn to see, appreciate, and transmit the light that others have already lit. If criticism is conservative, science, despite its equally rigorous requirements for training, is progressive. This ideological opposition underlies the conditions that C. P. Snow (1959) lamented, the supposed mutual ignorance of "the two cultures" of the sciences and the humanities.

Aristotle, of course, did not see the world so divided. For him, "artists" (*technitês*) were those who know not only that something is but why it is (1951, 68–9), which is to say, they were what we would call scientists. *Technitês* practice *technē,* which Aristotle defined as "a trained disposition to make in accordance with correct calculation" (1951, 227). *Technē,* which Aristotle used where we would use "art" (as in "the art of medicine" as well as "the art of poetry"), is the root of our word "technology". What we would call "the liberal arts" required training with technology, be that technology the mastery of a lute for the production of music or the mastery of geometry for the parcelling of land.

Humanities departments today, too, focus on technology, although we are usually unaware of that fact. English departments focus on the mastery of the English language per se and of rhetoric in particular. The objects of criticism in English departments are examples of art created with the technology called English. For its native speakers, English is probably the most significant technology of all for it is central to the organization of labour and the conduct of social life. This technology, English, is to language as monkey wrench is to wrench. The other language departments also define themselves centrally around technologies: pipe wrenches, allen wrenches, and so on; French, Japanese, and so on. Art history departments spring from the technologies of painting and sculpture, film departments from the evolving technologies of cinema, history departments from the scrutiny of the documents and material products left by the work of others. Once we recognize this technological basis for current humanities departments, we should understand that the challenge of digital media to the traditional humanities is not an assault but a source for revitalization.

Professing in the digital infosphere

Once upon a time, Edward Gibbon's monumental *History of the Decline and Fall of the Roman Empire* (1776–1788) was read as history. Now, given that our modern sense of ancient Rome has long ago supplanted Gibbon's positions, *Decline and Fall* is equally monumental, but in a very different sense: the book functions as an historical document in a past debate about English imperialism. It is also, though, "literature," that is, a work of language read in significant part for the interest we have in how it uses language itself. Is Gibbon's a good book? The answer to that depends on how you judge its audience and purpose. As Lewis Carroll's Alice says of a changeling in her arms "If it had grown up . . . it would have made a dreadfully ugly child; but it makes rather a handsome pig" (1960, 87). What was true in Wonderland is true in our classrooms. When we say that a sentence of prose or a line of poetry is "good," we are really saying that it is somehow effective for a certain audience in achieving a certain purpose.

Audience and purpose should often be thought of as a single concept, much as we do ham-and-eggs or ham-and-rye. Yes, there is such a thing as ham, but ham-and-eggs is a breakfast dish, not a pair of breakfast dishes, and ham-and-rye is a sandwich, not a two-course meal. Knowing that your audience is your mother, we tell our students, you write one way if you're asking for money, another if you're asking for romantic advice. Knowing that you're asking for money, you write one way if you're writing to your mother, another if you're writing to your best friend. There is no fixed rule about how to write to a given audience or for a given purpose; we must always explore, and teach our students to explore, audience-and-purpose. If we are "to learn . . . the best that is known and thought in the world," we must first recognize it, which means understanding what we encounter in terms of audience-and-purpose. If we are to "propagate" it, we must be able to judge our own speech and writing in terms of audience-and-purpose. The fundamental centrality of audience-and-purpose both to criticism and to composition underlies the typical commingling of these functions not only in language departments but in every course in language departments.

Analysis of audience-and-purpose underlies all genre criticism, say in considering children's literature versus adult philosophical tale. *Alice in Wonderland* is much too dark for most parents to wish their eight-year-olds to read it, but, fortunately, most eight-year-olds miss the ubiquitous death jokes because those jokes are much too subtle for them. As a dark work for children, Alice is weak; as a romp, it is strong. Yet as a dark work for adults it is strong; as a romp, it is juvenile.

The centrality of audience-and-purpose is hardly restricted to language. When I judge a dictionary as good or bad, I have very limited demands for narrative interest. When I judge an essay as good or bad, I have very limited demands for graphic interest. And yet I could have such demands, and if the essay were presented through the technologies of Web publication, I would. Once upon a time, we took medium for granted. We took even genre for granted. But we should not. Just as challenging fantasies like *Alice* foreground written genre, challenging technologies like the Web foreground medium itself. As teachers, we need to explore medium today whether once we ignored it or not.

In a similar way, once upon a time, we didn't need to worry much about evaluating sources. If a book had the right imprimatur, we could rely on the publisher's management of the peer review process as guarantor of value. But our students now turn first to the Web, and perhaps there is no second turn if we don't demand it. On the Web there is such a thing as peer review—sometimes—but what do we mean by imprimatur? We need to teach the evaluation of sources, and some have offered us schemes and resources to do so (Engle 1996). Again, the new media vivify what once was too often unconscious but always necessary. To profess in the digital infosphere requires working consciously with concepts of resource evaluation.

In traditional courses, we always had to teach the difference between foraging and formation, between merely collecting stuff in a library and creating an original piece of writing that reflects some original thought. But with foraging so easy that students may be seduced into believing that they are masters of a subject simply because they have found six web sites that mention it, we must more urgently clarify the need to make something of what they have found. Not only must they evaluate their materials for authority—deciding whether or not to trust the site—but for sensibleness. Students must understand that they need to become critics of everything they read, a fact that makes the profession of literature all the more crucial, and that valorizing of criticism does not mean only to learn the best that has been known and thought but to strive to contribute to the store of such excellences. Fortunately, the mechanics of contribution are easy in the new infosphere. Tell your students that their work must be on the Web and suddenly they blink in the bright light of potential scrutiny. They tend to develop a sense of imminent audience.

And yet, despite this sense, the ease of copy-and-paste makes plagiarism ever more seductive. What can we do? Well, copy-and-paste is a useful strategy for the plagiarist at only one stage of composition, the last stage. But the best work goes through many stages. We can ask for a proposal for a paper topic. We can comment on that proposal and ask for it to be refined. We can ask for a draft of a thesis statement. We can ask for students to exchange and comment on drafts of each other's work, comments to which we respond. We can build into our syllabi the opportunity for evaluated revision. If we involve ourselves with our students at many stages of their efforts, the strategy of plagiarism becomes more trouble than it could possibly be worth, but that is not why we should be involved. Protecting the academy against the misdemeanours of the infosphere requires that we be the better teachers we always wanted to be.

The production of student work within an active social framework, be it simply the frequent consultation with the instructor or the more general collaboration with instructor and fellow students, highlights the fact that all cultural production is done for some audience-and-purpose. Networked computing makes possible the creation and strengthening of social ties. One consequence of this is that humanities education is less a matter of setting an assignment and waiting for a product than it is a matter of ongoing consultation and the evolution of original contributions. We all have access to the texts now, and we can all use a word processor to find all occurrences of any given word in the text. What is important is thinking of which words to find, thinking about what they mean in their various contexts, countering the weak arguments of others, constructing new arguments to enrich the lives of . . . whom? The instructor? Perhaps. Fellow students? I should hope so. The whole world? Why not? And thus the distinction between teaching and research erodes.

Renewing practice

In a world in which teaching and research often become indistinguishable, we find happily that we can demand real creativity from our students. But to create means to use tools, even if only the traditional tools that define humanities departments (language, art, and so on). In the current era, there are more tools to use, and we need either to teach them or have them taught to our students. Visual literacy, for example, would seem to be a necessary skill, especially since it would encourage us to become instrumentally conversant with and communicatively aware of technology. Media—digital images, for example—are after all worked by technologies—Photoshop, for example. We use a given media as we choose. There is no such thing as a photograph or a manipulation of a photograph without some style, just as there is no such thing as a text without some style. Style has meaning. Digital media make that vivid. If we need to teach about linguistic style to help our students produce the best possible prose essays, we need to teach about visual style—and teach the tools to manipulate visual style—to help our students produce the best possible multimedia essays. In our infosphere, one in which our students gain much of their knowledge in a multimedia environment, that move from prose essays to multimedia essays seems imperative.

Fortunately, the move to multimedia highlights many issues—in this case style—that we have always striven to make conscious for our students. Even if we do not intend to encourage the production of multimedia essays or teach the use of Photoshop, in today's infosphere, we should be able to use multimedia to foreground issues, like the significance and malleability of style, that have always mattered to us.

Plato condemned the technology of writing in the *Phaedrus*: "If men learn this, it will implant forgetfulness in their souls" (1961, 275a). The crucial problem, though, was not merely a weakening of memory but of community and the discourse that community supports. You can interrogate a speaker but not a written text. Did Plato, then, eschew writing? Not at all. His solution was to write dialogues. He used writing to give us the experience of the verbal world. We need to follow Plato's lead, to think about media, highlight media for our students, and use diverse media to create the communicative environment we desire. To do so means, of course, teaching the tools.

Tufte's critique of PowerPoint (2003), that it tends to reduce all thought to bullet points, is fair. But it is no more dispositive than Plato's critique of writing. In English 415/516, we have every student create a free-standing PowerPoint presentation, that is, a work meant for a user to use alone, not something meant as part of the creator's live performance before a group. The subject of this free-standing presentation is the humanistic implications of some technology, the technology in question to be decided upon by extensive consultation with the instructor. The subject matter focuses the students on technology and the experience of composing in a new medium—one in which students must ask themselves when to use pictures, sounds, animations, slide transitions, object builds, hyperlinks, and so on—makes them exquisitely aware that at least this technology is more than just an extension, as heels make us a tad taller, but rather a transformative reality, like the automobile. Students inevitably infer from the experience of pursuing this assignment that all technologies, certainly writing and language and perhaps even the heels that make some politicians seem taller, are potentially transformative. The assignment conveys an argument by experience. In the digital infosphere, we can pick the tools and tasks we set before students to help them learn more deeply than they otherwise might.[1]

The modern infosphere tends to require and reward collaboration. Most of the sites our students visit are created by teams. We want our students to publish, too, since the digital infosphere makes publication readily possible. We want our students to sharpen their work under the felt pressure of a wide audience. But in a world with so many good sites, how can student work hold its own? The answer, of course, is collaboration.

Many instructors have required collaboration for decades, but most instructors have also found that group work leads to common student complaints, typically about the failures of some to contribute reliably or about the difficulty of working out ways to collaborate. These are usually valid complaints and they arise from the simple fact that collaboration—which usually goes untaught—is no simple skill. The answer is to teach it.

One of the online resources associated with my course web sites is Collaboration Tools (http://www.umich.edu/%7Emmx/collaboration_tools.htm.). This page lists a number of digital tools to expedite collaboration, from editorial tracking tools in Word to setting permissions for document folders in cyberspace; however, what is most important is the sequenced list of social techniques under "Group Organization," a list that tells students what they need to do in order to collaborate and the order in which they need to begin doing those things. The tools and the techniques form a set that works, although I never merely point the students to the web page. Instead, I discuss each item with them, listen to their ideas, and give them a model of collaborating in class. And I require that the tools be used in ways that open up the community. Each group, for example, is required to establish a mail group for one-to-many communication, and I require that I be made of member of each such mail group. The students come to know quickly that I may be aware, at some time or other, of anything they say electronically at any time. I may make a strategic intervention, but usually I am silent. Nonetheless, the felt presence of the instructor keeps most people on their toes and also makes palpable a sense of audience, a fact that energizes and focuses their minds and work.

In both these courses, the collaboration requires a shared set of basic learning across all group members and specialization by need, talent, and desire for individuals. Thus graduate students from our School of Information may be led by junior English majors in interpreting texts while graduate students in English are helped by sophomore Computer Engineering students to make an interactive image into a complex but meaningful menu for a group site. And all can communicate with each other well because all share at least rudimentary knowledge on both the humanities and technology sides of the course. We insure that shared minimum knowledge in both the humanities and digital technology by teaching these things. The assignments that accomplish this are on the course web sites. The students, as well as the general public, can access them at any time.

In the digital infosphere, time is untethered. No suns rise or set, web pages aren't numbered, homework on the Web doesn't have to go away when the semester ends. The combination of this timelessness with the inherent malleability of digital work means that what one student group does today another may want to augment or modify tomorrow. To this end, English 414 requires each student to sign a letter assigning certain aspects of copyright to me as a representative of the University of Michigan so that other students can work with those student-made materials. (An example of such a letter, of course, is online in the English 414 web site.) Whether or not this assignment is ever exploited, the sense that the work upon which they are embarking is potentially lasting and important focuses the mind wonderfully.

Of course, in the digital infosphere, copyright is important not only for student work but for anything one might forage and then form. So copyright needs to be taught explicitly. Again, the course web sites have links (under Supplementary Materials) to sites dealing with that subject, but the instructor must make the issue explicit and discuss copyright. Copyright law is based on property law while plagiarism is a variety of fraud. One needs to make these differences clear, a much harder task than we used to have in traditional literature courses. How do you teach it? First, you need to learn it. The Web can help. But first comes the desire.

The digital infosphere puts enormous demands on humanities instructors, crying out for us to teach tools (like Photoshop) we may not ourselves yet know, subjects (like law) we may not ourselves yet know, ideas (like visual style) we may not ourselves yet know. But if we answer the cry, we will find ourselves more deeply learned in a wider sense of the humanities and more able to teach in more powerful ways than traditional practice made possible.

Renewing theory

Practical changes in pedagogy are often inseparable from changes in theory. That is equally true in moving humanities teaching into the digital infosphere. Let me offer here just two examples.

In the study of phonology, linguists seek "minimal contrastive pairs." In English the difference between "k" and "q" makes no difference. English speakers hear the "k" in "kaffiyeh" and the "q" in "Iraq" as the same sound. But Arabic speakers hear two different sounds, which is why we transliterate Arabic using two different Roman letters for sounds we hear as one. We can learn to hear the difference between "k" and "q" by saying the word "kick" slowly aloud. Feel your tongue. It touches your hard palate with the first "k" and your soft palate with the second. Do it again and hear the different sounds the different techniques produce. That is the difference between "k" and "q" in Arabic. To Arab speakers, "qaffiyeh" and "kaffiyeh" would be two different words.

When we teach poetry and say that "My love is like a red, red rose" is a "good" line, we are implicitly suggesting that it is better than "My love is like a fresh red rose" or "My love is like a new red rose," not that it is better or worse than "To be or not to be." What we should be doing explicitly when we seek to evaluate poetry, but usually do only implicitly, is construct hypothetical minimal contrastive pairs. Let me make two points about this exercise. First, we could become better critics, and teach our students to be better critics, if they made such imaginative comparisons explicitly, not only about poetry but about all objects of humanistic study. Second, even having the pairs, the judgment of "goodness" still depends on a sense of audience-and-purpose. If Burns had been discussing his passion for gardening with an old school chum, the line would strike us quite differently than it does as a lyric about eros. One of the student groups in an early offering of English 415/516 created a web site about Blake's *Songs of Innocence and Experience* that includes what one now might think of as a digital version of refrigerator magnet poetry. The user could add three new words to a Blake poem and rearrange those and all others to see if Blake's lines were really best. For copyright reasons, this site is no longer on the Web, but it has been honoured by inclusion in the US National Archives as one of the outstanding multimedia works of its year. Using digital media created an attractive domain in which users, who typically don't bother making minimal contrastive pairs explicit, readily spend half an hour consciously playing with them. The general power of the technique of seeking minimal contrastive pairs, and the ability of digital technology to foreground that technique, is one point of theoretical renewal.

Will Eisner, one of America's great comic book artists, divided images into two sorts, "illustrations" and "visuals." An illustration merely shows us a picture of what the text describes, like a photo of Lincoln next to his entry in a dictionary. A visual is a graphic that participates actively in the telling of the story, the way well-written children's books rely on pictures (Eisner 1985, 127–8). To this distinction, I would add another, "decoration." Walter Crane (1972, 17), in the first theoretical discussion of graphics in books, writes that, "In a journey through a book it is pleasant to reach the oasis of a picture or an ornament, to sit awhile under the palms, to let our thoughts unburdened stray, to drink of other intellectual waters, and to see the ideas we have been pursuing, perchance, reflected in them." As we go along a continuum from decoration, which is a pretty place to rest and stops the narrative entirely, to illustration, to visual, which forces the narrative along, we move from less to more information density. Is information density a good thing? Sometimes, as when we praise the aesthetic virtue of "economy." But if a text is so economical, a picture so demandingly new in the text, that we can't follow the reading, the equally valuable virtue of "unity" has been compromised. Since the ancients, the productive tension among economy, unity, and variety has been well known, in those or equivalent terms. Today we can see that those criteria extend beyond language and raise questions about the local wisdom in a given work of attempting to involve a user in more or less dense information.

These two theoretical notions that grow from and inform humanities teaching in the digital infosphere—the utility of seeking minimal contrastive pairs and the explicit recognition in the decoration-illustration-visual continuum of the significance of information density—are but two of many. The erosion of the school-bred distinction between teaching and research is clear; the erosion of the unity of the author is clear; the erosion of the passivity of the user of humanities is clear. And in that lies the wonder. For thanks to the digital infosphere, which we now inhabit whether or not we consciously confront it in our classrooms, the humanities are no longer a separate culture but interwoven as they once had always been in the fabric of every truly liberal—and liberating—education.

Note

1. For those interested in seeing the syllabi for the courses in question, please find links to them on my web site (http://www-personal.umich.edu/~esrabkin/), which also contains a link to Selected Student Humanities InfoTech Coursework, a menu of real examples of free-standing PowerPoint presentations by individual students and web sites by student groups.

11
The Rhetoric of New Media: Teaching a Rhetoric of Hypertext

Jeff Rice

The textbook: a pedagogical moment

When we speak about new media, we must also speak about how we study and teach new media. I want to begin such a discussion with a quotation from Robert Atwan's *Convergences* (2002, xxxviii), a popular composition textbook whose thematic focus is new media.

> We must develop an awareness of how media penetrates nearly everything we see and hear. We need to understand how one or another medium is always present, molding and filtering expression, even when it pretends to be invisible. Even when it disguises itself as reality.

Atwan's remarks are informative for any number of reasons, but the most impressive meaning for me resides in its concentration on reading, not writing. As an introductory statement in a writing textbook, Atwan's comments signify comprehension as the most vital component of new media study. Awareness of media construction, the quotation suggests, produces critical thinking. It's hard to disagree with such sentiment, yet I am left wondering why a writing textbook interested in teaching new media stresses reading, and not writing itself?

There is a generalizeable lesson to be learned from a textbook like *Convergences*. I don't mean to single out this specific textbook as an exception. Instead, *Convergences* allows me to refer to how textbook production, in general, currently treats new media. While Atwan cautions students to recognize the constructed nature of media representation, his textbook—even though it is a writing textbook—does not address how students themselves can write in media environments. Even though Atwan's advice is sandwiched between various images of web pages, advertisements, photographs, and other new media representations, no part of the textbook considers how such productions are rhetorically constructed. Are students to believe that these works are "written" elsewhere? Are they to believe that while they may eventually become empowered to decipher new media representations (with the assistance of the textbook), they will not be able to construct their own representations? Why does a new media textbook teach a print-based pedagogy (hermeneutical reading)? Why does it minimize issues of new media rhetorics; in other words, why doesn't it teach a new media rhetorical approach to expression?

The origins of the composition textbook, of course, reside in rhetoric. The nineteenth-century shift from oral methods of information delivery and education dominant in the rhetorical tradition eventually yielded to print culture's needs for students who could write, not just speak, persuasively. The subsequent emergence of the composition textbook as a pedagogical tool, according to Robert Connors (2003, 100–01), stems from the shift in the classroom from oral instruction (lecture) to print assessment (exercises performed through writing). Around the mid to late 1800s, rhetorical treatises, the dominant form of rhetorical instruction, began including question and answer sections as well as drills and exercises with each text's chapters. Whereas students had once merely copied down a lecture and then recited it by heart, the new rhetoric demanded that students formulate their own compositions by learning from previous rhetorical examples how such compositions are constructed.

> The conception grew that one learns to write by consciously learning ideas about writing and then practicing the application of these ideas. The story of the growth of composition textbooks is the story of the abstract and theoretical rhetoric that was the legacy of the treatise forcing itself into realms of skill development not easily comfortable to it. (Connors 2003, 106)

The invention of the composition textbook answered a late nineteenth-century need for instruction relevant to changes in communication.

This historical moment should not be lost on those concerned with contemporary writing instruction and the emergence of new media. Connors' work can be applied to our current situation regarding writing, pedagogy, and new media. When I encounter a contemporary textbook designed for new media, like *Convergences*, I question whether our pedagogical practices address present communicative changes. Even though a textbook like *Convergences* includes a variety of visual examples representative of new media, its pedagogy does not assist students eager to produce such work as well.

Lest we think that all textbooks ignore new media production, textbook publisher catalogues do often display titles directed towards the most popular of all new media forms, the World Wide Web. Unlike the rhetorical approach *Convergences* promises but is unable to deliver through actual new media instruction, these titles promote instrumental (that is, how to) training regarding web writing. Such instruction includes how to make links, tables, frames, or how to design "usable" sites. Margaret Batschelet's *Web Writing/Web Designing* (2001) and Johndan Johnson Eilola's *Designing Effective Websites* (2002) are two such texts produced by textbook publishers, both useful for their specific purposes, but limited in terms of rhetorical production.

Just as nineteenth-century pedagogy insisted on developing a print-based educational process that encouraged the production of textbooks geared towards how-to's (how to form a sentence, how to punctuate, how to choose the right word), so too have new media texts identified their purpose for writers eager to participate in online culture. In many ways, texts like Batschelet's and Johnson-Eilola's are the contemporary equivalent of William Strunk and E. B. White's popular *The Elements of Style* (1979). While no longer a dominant text in composition studies, the 1918 first edition of *The Elements of Style* was itself a response to the changing writing curriculum. Its appearance complemented writing programmes' efforts to solidify assessment of print-based skills (spelling and grammar) as opposed to oral delivery. The mythic role of Strunk and White's text in writing instruction has been largely upheld through its how-to treatment of rhetorical application: when to use "active" vs. "passive" voice, or "that" vs. "which," or "can" as opposed to "may." Like a text that teaches how to make web pages, *The Elements of Style* is a how-to book, a practical approach to a once-new media form, print.

The rhetoric of hypertext: a beginning

To begin, I want to focus on one form of new media, hypertext, both because of its familiarity to us via the Web and because of its complicated role in composition studies where a subarea of composition, computers and writing, encourages teaching hypertext in the writing classroom. Much early enthusiasm for hypertext exists in early 1990s scholarship which expressed fascination with its linking because it supposedly mimics associative thinking. Often heralded by critics of the early 1990s as "natural," "nonstandard," and "outside of the ideological constraints" imposed by hierarchical thinking, these writers (such as Ted Nelson, Stuart Moulthrop, Nancy Kaplan, George Landow, Michael Joyce, and Jay David Bolter) romanticized the new technology. Such views left unchallenged the ways in which hypertext authoring systems used primarily to teach linking (like HyperCard or Storyspace) passed unchallenged for how they produce, or don't produce, electronic rhetorics. In addition, those writers dominant in early hypertext scholarship considered hypertext a literary form—partly due to the preference for Storyspace, a preWeb stand-alone hypertext authoring system responsible for the publication of the medium's most canonical literary works and based primarily on creating alternative forms of fiction through complex interlinking.

Technology, however, has outgrown the vision of hypertext as only linking. Indeed, web-based developments over the last few years have shifted Ted Nelson's original definition of hypertext as nonsequential writing to a more multimedia method of expression centred around HTML tags. What Nelson hypothesized in the early 1960s has changed due to technological improvements and to such additions to hypertext as Perl, DHTML, Flash, and XHTML, among others. The link remains important to hypertext. So do other factors as well: hypertext remains nonsequential as well as a number of other things. As Mathew Kirschenbaum (2000, 121) writes, we need to ask "whether hypertext embodies something more—aesthetically, conceptually, or computationally—than just the mechanical process of linking."

My challenge is to design a method for teaching hypertext whose focus involves more than linking. In other words, I want to work towards creating a rhetoric of hypertext, a practice which addresses both the rhetorical and how-to issues relevant to new media production. As a writing instructor, the temptation is to first rely upon current "best" practices associated with rhetoric and hypertextual instruction in order to situate my work within the relevant pedagogy in the field. The most typical place to find such practices is in the professional writing programmes. Professional writing's attraction to hypertext stems from the discipline's recognition that contemporary professional communication takes place or will take place in new media environments. Professional writing's understanding of hypertext, however, sees "usability" as the main feature of hypertextual writing. Influenced by design experts like Jakob Nielsen or Patrick Lynch, professional writing identifies and teaches a business approach to hypertextual instruction, one in which writerly purpose often translates as "commercial objectives" and where audience means "client." The rhetoric of hypertext for these writers, and those writing programmes that advertently or inadvertently base their curriculum on such thinking, includes the creation of sites which are "easy to use" and "manageable." HTML-inspired assignments in such courses ask students to put résumés online, prepare software documentation, and design web sites for imaginary or real corporations. In short, this kind of instruction treats hypertext rhetoric as business discourse.

Professional writing has been useful to English studies for a variety of reasons. My disagreement with this business-oriented approach is that its focus is less on rhetoric and more on commerce. Rather than develop a pedagogy that prepares students for only professional workplace communication, I want students to address the new rhetorical challenges hypertext presents so that they can construct their own meanings for a variety of situations and contexts and not just those relevant to trade and finance.

A rhetoric of hypertext

Instead of basing my inquiry on professional writing, I take the first step towards developing a rhetoric of hypertext by borrowing the momentum initiated by two previous works related to hypertext and rhetoric: George Landow's "Rhetoric of Hypermedia: Some Rules for Authors" (1991) and John Palattella's "Formatting Patrimony: the Rhetoric of Hypertext" (1995). These essays mark initial forays into rhetorical issues perceived to be associated with hypertextual writing. While both are instructive for my purposes, neither essay constructs a contemporary rhetoric for hypertextual writing. Palattella discusses writings *about* hypertext, not hypertext itself, and he uncovers a rhetoric consistent in theoretical writings (that is, the language such authors use when discussing hypertext). Landow maps out a rhetoric and states that his rhetoric will go beyond the link in developing a hypermedia rhetoric. Landow's focus is navigation, a system for hypertextual reading, which ends up closely aligned with linking. Landow's attempt to codify this supposed rhetoric translates into a set of ambiguous rules ("links should stimulate a reader to think and explore" [1991, 86]) centred on how one creates links, what one links to, and where those links appear on a given document. In all, Landow's work teaches me that linking cannot be ignored if I am to fashion a rhetoric of hypertext, but that it also cannot be the sole issue involved.

A more significant lesson for my project comes from Nicholas Burbules's "Rhetorics of the Web" (1998). Burbules points to the link as a rhetorical application, paralleling its discursive significance with various rhetorical tropes and figures: metaphor, metonymy, and hyperbole (among others). "I want to show links as rhetorical moves," Burbules (117) writes, "that can be evaluated and questioned for their relevance. They imply choices; they reveal assumptions; they have effects." For each rhetorical trope or figure, Burbules identifies a similar application for web writing. In addition to his categorization of linking possibilities, Burbules (118–19) closes with a call for understanding how web writing functions:

> Just as specialists in other fields (from poetry to acting to political speech writing) can be the sharpest critics of other practitioners because they know the conventions, tricks, and moves that establish a sense of style and elicit particular responses from an audience, so also should hyperreaders (whether or not they actually design/author material for the Web themselves) know what goes into selecting material for a page, making links, organizing a cluster of separate pages into a hyperlinked Web site, and so forth. The more that one is aware of *how* this is done, the more one can be aware *that* it was done and that it *could* have been done otherwise.

Burbules shows me how to conceive a hypertextual rhetoric by teaching me to conceptualize such a practice with traditional rhetorical instruction. Yet his work also brings me full circle to the textbook I began with. Burbules, like Atwan, deemphasizes the writing process in favour of reading. Once again, "awareness" is reduced to a reading practice. The effectiveness of critical awareness that Burbules underscores, I contend, means little until students become producers of web-written information and not just "hyperreaders." To engage in a pedagogy of writing instruction for the Web, I also borrow Burbules' requirement that students understand how a site is constructed, but I do so in order to teach students how to construct sites rhetorically themselves.

What is a hypertextual rhetoric?

To create a rhetoric of hypertext, I return to Aristotle's *The Rhetoric*, a text which educated Greek students to learn how to produce discourse for various contexts and audiences. Aristotle teaches me that rhetoric is about production—that is, producing texts. To create a rhetoric of hypertext, I need to update Aristotle's text with acknowledgment of technological innovations in circulation since ancient Greece. Composition studies, and rhetorical studies in general, often treat Aristotle's work as if communicative technologies have not altered the ways we construct discourse. We need a hypertextual version of Aristotle's work, a new media rhetoric that will teach us ways to write for the Web.

In order to begin such a discussion, I draw upon three major new media thinkers who theorize the effects of technology on rhetoric. Then I will consider how my own teaching learns from these writers and works towards teaching rhetoric and composition from a new media perspective. My brief outline of these three theorists, then, acts as a preliminary recipe, a set of instructions I borrow from each theorist in order to organize a rhetoric of new media, and more particularly, for hypertext. From each writer, I learn an application relevant towards creating a rhetoric of hypertext.

McLuhan. Prior to hypertext, Marshall McLuhan theorized the impact of new media on composing practices. McLuhan recognized that just as the Gutenberg press changed the nature of rhetorical production through page design, automation, and a general literacy shift, so too will new media affect how we construct discourse. Noting the way new media (film, television, and to some extent computing) patterns information, McLuhan considered collage an appropriate rhetorical feature of new media writing. Thus, he chose to mimic the collagist process by constructing the majority of his books as collages, which juxtapose other writings with his own insight. New media, McLuhan taught, create interdisciplinary interaction (and thus, discursive interaction) in complex ways best demonstrated by the collagist juxtaposition. "The instantaneous world of electronic informational media involves all of us, all at once," McLuhan writes (1996, 53). From McLuhan, I borrow an important insight regarding new media production: juxtapose.

Manovich. Lev Manovich is one of the first contemporary theorists to consider how new media functions as a language by examining "the emergent conventions, recurrent design patterns, and key forms of new media" (2000, 12). By inventorying web sites, computer games, and software, Manovich (27–48) finds five principles relevant to new media: "numerical representation," "modularity," "automation," "variability," and "transcoding." By looking at interfaces, operations, and database construction, Manovich attempts to construct a new rhetoric for media work. His rationale stems from "observations" regarding current media formations. Manovich teaches us to observe and inventory the media we encounter and communicate with in order to understand how it functions. "To develop a new aesthetics of new media," Manovich (314) writes, "we should pay as much attention to cultural history as to the computer's unique new possibilities to generate, organise, manipulate, and distribute data." Manovich's lesson for a hypertextual rhetoric: inventory structure.

Ulmer. In *Heuretics: the Logic of Invention* (1994), Gregory Ulmer attempts to invent a hyperrhetoric. "What will research be like in an electronic apparatus?," he asks (32). His response is chorography, an update of the Aristotelian *topos*. Whereas the *topois* were fixed places of argumentation, chorography (based on Plato's *chora*) is dynamic. Ulmer offers a set of instructions for how to be a chorographer: "do not choose between the different meanings of key terms, but compose by using all the meanings" (48). Concepts, words, and, places, all possess multiple meanings and associations. When brought together in unfamiliar ways, these meanings will produce new ideas and thus stimulate invention. For Ulmer, chorography is a method conducive to writing in new media. Print's fixation on the page made it a suitable medium for the *topos*. The ability to juxtapose media, to work with imagery and links, to construct video representations, works to *chora*'s associative nature. Because of new media's affinity for association, Ulmer's rhetoric asks that patterns guide composition in new media practices.

What I learn from these three writers is how to think of new media in terms of rhetorical production; each teaches a "method" for producing discourse. These writers generalize rhetorical approaches to creating new media for most electronic writing. McLuhan teaches juxtaposition, Manovich the inventory, and Ulmer pattern formation. Together, we might label them all as a rhetoric of hypertext. But I'm not ready for that just yet because missing from such a definition is the student writer herself. Through my formulation of a rhetoric of hypertext, I don't want to create a rhetoric absent of student input; in other words, I want to avoid creating another set of standardized instructions students follow (like the instructional how-to). Such has been the textbook model whose emphasis is on forming topic sentences, outlining, paragraph development, and using research resources in very specified ways. Such methods have proven useful to the institutionalization of print discourse. The novelty of hypertext (an eleven-year-old medium for most users) means the medium is not yet ready for such permanence.

This theoretical recipe I've constructed creates the initial groundwork for developing my rhetoric, but I must now shift attention to the classroom where students work from this introduction as they fashion their own rhetorical instructions.

The student writer

Aristotle's construction of a rhetoric, like Manovich's, requires that he inventories various areas of discourse as well as various situations where discourse occurs, and offers methods for speaking to each situation or audience. A rhetoric of hypertext must do the same. Unfortunately, there exist few (if any) contemporary sources for constructing a hypertextual rhetoric. Regarding writing instruction, courses that examine (according to their course titles) "digital rhetorics," or "electronic rhetorics," typically focus on nonrhetorical issues: community, gender, class, economics, race, and power. While each area maintains its own internal rhetoric (how a specific community communicates online, for example), these courses don't ask students to examine the rhetorical issues at stake in online writing so that they, too, can produce electronic texts. Indeed, these kinds of new media courses follow Burbules's or Atwan's interest in teaching students to be critical readers of new media, but not critical writers.

Contemporary handbooks of writing (Aristotle's rhetoric is, after all, a handbook) like *Convergences* fail to offer instruction regarding how to write electronically. They teach how to write about, not with, new media. In *Convergences*, Atwan (2002, xxxix) tells students, "you will be asked to examine all kinds of written, oral, and visual expressions as though they are 'texts' to read—and then you will be asked to write about them."[1] At what point, a new media rhetorician might wonder, does the student get to *write* like the new media text and not just *about* the text in question. Such was Aristotle's critique early in *The Rhetoric*: existing handbooks on speech explained much about speech, but not how to form persuasive speech itself; Aristotle sought to change this form of pedagogy by introducing what he called "artistic proofs" (1960, 3), the enthymeme, a practice one could engage in as well. This, then, is the continuing paradox in which students study new media but don't know how to write for new media, much like what McLuhan (1964, viii–ix) noted in the 1960s:

> At school, however, [students encounter] a situation organised by means of classified information. The subjects are unrelated. They are visually conceived in terms of a blueprint. The student can find no possible means of involvement for himself, nor can he discover how the educational scene relates to the mythic world of electronically processed data and experience he takes for granted.

When teaching new media, therefore, the challenge is to forge a relationship between object of study (the Web) and the student's discursive experience (how to write for the Web). The users of new technology eventually invent its rhetorical applications. Aristotle and Cicero didn't invent speech but devised ways to speak persuasively; Sergei Eisenstein didn't invent film, but constructed intellectual montage. Our students are the new media writers; they use (or will use) the Web in various ways for expression. In turn, they will invent its rhetoric.

Thus the problem for the contemporary writing classroom: how to invent a practice in order to write for the Web? The solution: invent a theory of web writing. The question which prompts the task is one of the apparatus—how to create a method appropriate to the institution where the method will be demonstrated. Thus, the first year writing course marks an excellent place to try out the experiment because first year composition has long been the American academy's place for institutionalizing writing. Beginning with Harvard's 1885 creation of English A, the first composition course, first year writing serves as a focal point for introducing students to developing and establishing discursive practices. The example I present, therefore, is situated in the first year writing course.

The handbook: another pedagogical moment

Since students must purchase and study writing handbooks in their writing classes, the most appropriate medium for inventing a theory of web writing is the handbook. The purpose of creating a handbook is both expository and pedagogical: the handbook acts as a descriptive set of instructions (this is what it means) and as a tool for teaching (here's how to do it as well). In addition, the tradition of using handbooks for pedagogical purposes extends from Aristotle's *The Rhetoric* (a handbook) to contemporary how-to's to the dominant presence of writing handbooks in most composition courses. Consequently, the handbook is a recognizable model for an assignment that asks students to create their own rhetoric for web writing. To invent a rhetoric of hypertext, students create handbooks which will outline a theory of web writing.

The instructions for students are as follows:

> Construct a theory for hypertext writing. To accomplish this you must spend time surfing the web, evaluating web sites, documenting your research. While the Web is not an all-inclusive reservoir of information, we will treat it as a library and use its resources to develop a theory for hypertext writing.

The assignment asks students to work with the following instructions in order to accomplish this task.[2]

> *Inventory*:
> Make a list of every site you visit.

In addition to mainstream, commercial web sites that merely reproduce print and television formats (i.e. cnn.com, espn.com, rollingstone.com), search out the unique, the strange, the bizarre. Visit the personal and the artistic, too: individual homepages, adaweb, rhizome, trashconnection, and others.

Outline the functionality of the sites.

Look for similarities between the sites you visit. How is information classified? How is it presented? What is the importance of graphics? Of javascripts? Of frames?

Make detailed notes of all of these observations.

Pattern:
Go through your notes and indicate patterns common to all of the sites you observed. Use these similarities, these patterns, to construct your theory. The rule of thumb is that if you see the same pattern reoccurring, there is a theory to be found there.

Present your findings in a hypertext document. Create your site based on the guidelines you develop from your web research.

Juxtapose:
Think of how you can use the theory you develop to design your project. Shouldn't your project adhere to the theory you are outlining? If you find "interaction" to be important in web writing, for example, shouldn't your site be interactive in some way? If you think interlinking is part of web writing, shouldn't that definition be interlinked? If you note that nonlinearity is a rhetorical principle of web writing, then shouldn't that definition be itself nonlinear? In other words, shouldn't you juxtapose your theoretical point with its example? As you write your theory for web writing, ask if your definitions are reproducible? Can someone else use your theory to write for the Web?

The benefit of such an assignment is that it allows the student to create a workable theory from which not only this assignment can be performed, but future hypertext assignments can be as well. The student, in other words, writes her own rhetoric. In the end, students develop handbooks that go beyond the link, which find framing, juxtaposition, nonlinearity, interactivity, image mapping, visuality, and other points to be rhetorical principles. They test such principles in their own writing; and other students, reading each handbook off the Web, test the principles in their own work as well. A section on the importance of nonlinearity is itself nonlinear. A section that considers the effect imagemaps have on hypertext uses an imagemap to do so.

The purpose of this assignment is not to codify hypertext into a set standard. Such has been the guiding motive of both online and composition approaches to teaching hypertextual writing. Both professional organizations like the World Wide Web Consortium (W3C) and professional writing programs represent hypertextual writing as a set of standards already set in place. The project which asks students to develop a rhetoric of hypertext begins from a contrasting position; we are not ready to standardize a rhetoric of hypertext. Instead, we must continue to work towards understanding how discourse forms online. In turn, student projects don't homogenize hypertextual rhetoric through their analysis, but rather establish a working set of principles that can be extended or minimized depending on the students' rhetorical needs.

In courses like the one I've described, we work to understand hypertext as rhetoric so that we can be persuasive in this new medium. In other words, what I'm describing, and what is often missing in analysis of hypertext and new media in general, is pedagogy. More importantly, in this assignment, most of the pedagogy results from the writers' own work. The hermeneutical analysis of web sites (the Atwan and Burbules model) does not represent a pedagogical moment as much as an assessment moment (show us what you've learned). The pedagogical practice of invention (inventing a theory of web writing) depends on the student learning from her own and others' teaching.

In itself, the handbook does not establish a complete rhetorical treatise. What is most obviously missing is the critical application of the rhetoric it teaches. One might ask: how can the frame be used to critique racists practices in business; how can juxtaposition of image through the rollover produce persuasive reasoning regarding local political debate? How does nonlinearity create social critique? These questions are not immediately addressed by the handbook. What it does do, however, is it gives students the tools to move in this direction. A follow-up assignment to the handbook would be to address one of these kinds of issues through the terms previously defined.

I have moved then from the institutional moment of instruction—the media-based textbook—to the rhetorical tool of instruction—the handbook. In doing so, I ask that we refocus our attention to new media practices like hypertext by having students explore how such practices work in their own writing. The handbook is one option; I encourage others as well.

Notes

1. Ironically, this statement appears below a visual insert entitled "How to Erase a Skater." The visual example is from software documentation instructing users how to use that piece of software to produce a visual effect. The author's caption to the image does not ask the student to engage in similar work, only to recognize that images can be manipulated by others.
2. Most obviously missing from this breakdown of the assignment are the various things one does in a classroom to supplement the written instructions. Thus, I don't mention the conversations we have in class surrounding this assignment (which web sites to visit, examples of what we might identify as rhetorical in each web site, questions and answers students ask and receive, etc.). I also do not mention in this essay the weeks leading up to the assignment that include basic instruction in HTML. Readers of this assignment, therefore, should understand that instructors must supplement my example with additional inclass work.

12
Learning Secretary Hand: an Interactive Tutorial

Andrew Booth, David Lindley, and Oliver Pickering

The idea for the interactive program that has been developed at the University of Leeds for teaching secretary hand of the sixteenth and seventeenth centuries sprang into being in an early-morning moment, when a number of different thoughts crossed and coalesced. Its immediate context was the fact that the School of English was about to offer for the first time an MA in Renaissance English Literature, and we were considering what research skills it would be appropriate to include in the compulsory core module. This module was also to be made available to students beginning Ph.D. study in the area, and for both categories of student instruction in the reading of manuscripts was an obvious contender. Apart from the general principle that every serious student of the period should be able to handle some of the most important primary material, which survives only in handwritten form, the acquisition of basic palaeographical skills would extend the potential range of students' research, and, more immediately, would encourage them to make full use of the resources that the Brotherton Library could offer. The problem was how to approach the subject in the context of a short ten-week course, where many other topics must also be included. The learning of the distinctive and difficult forms of sixteenth and seventeenth-century secretary hand takes time and practice. The first we did not have, and the second is difficult to arrange. Traditionally palaeography has been taught either by using originals or photocopies of manuscripts, or else through anthologies of reproductions, which print transcriptions on a facing page (for example, Dawson and Kennedy-Skipton 1968; or Preston and Yeandle 1999). The drawbacks are obvious enough—getting access to originals or photocopies can be difficult, or impossible. The danger of the anthologies, useful though they are in many ways, is precisely that it is too easy when faced by difficulties in the script simply to crib from the transcription. In this context, then, what seemed necessary was some system that would enable students, after brief introduction to the basics, to practise transcription in their own unsupervised time.

David Lindley's original idea was simply to ask whether it might be possible to set up a computer program that offered a digitized photograph of a manuscript on one side of the screen, and on the other a window in which the student could attempt a transcription that the program could be asked to check at any time—if errors showed up the student would be able to return and try again until a final, correct version was achieved. It seemed important that any complete rendition of the manuscript should be hidden, so that the student would first have to struggle independently to make sense of the text. The University Library certainly had manuscripts that would serve as examples—if they were interested; the question was whether there was the technological expertise within the University to create an appropriate program.

Fortunately Oliver Pickering of the Library's Special Collections department was immediately interested in the idea, and by happy coincidence Andrew Booth, of the Flexible Learning Development Unit, though a scientist by training, had a strong interest in family history, and had consequently acquired skills in reading early hands. The team to put the idea into practice was thus assembled, and the Virtual Learning Environment used at Leeds, known as Bodington Common, provided the ideal site to deliver the program, closely modelled on David Lindley's original vision. The Bodington system offered a number of advantages. It is an environment where students encounter a variety of teaching materials, and therefore one with which they would already be familiar. Access to materials for any module can be restricted to the students registered for it, and it is easy to provide a "discussion" area, so that communication between tutors and students can be both immediate and confined to the group. This seemed particularly appropriate for work at an experimental stage. One of the virtues of delivery through the Bodington system, however, is that access to a correct transcription can be controlled by the tutor, so that students are not able to take short cuts but must, at least initially, experience the frustration of struggling to make sense of the documents before them.

I

From the beginning we were clear that this was not to be a program that aimed to give students an introduction to the detailed history of the development of handwriting in the period, nor one that attempted to equip them to become expert palaeographers. It was, and still is, intended to provide sufficient practice for postgraduate students to be able to confront original manuscripts with a basic working knowledge of the forms of secretary hand, and with sufficient confidence in their ability to read it to be able to use manuscript materials in their work. The original intention was that students would be introduced to the basics in a single classroom session within the Library, using paper-based materials, after which they would be expected to practise in their own time on the manuscripts digitized for the package. In choosing the manuscripts, therefore, we wanted to include a variety of secretary hands, including a fair number of mixed hands where the presence of later and more familiar letter forms would give students some initial confidence.

To start with—in summer 2000—we therefore chose fifteen short poems, or extracts from poems, all of the earlier seventeenth century (and illustrative therefore of easier rather than more difficult secretary hand). The items chosen all came from Leeds University Library's extensive collection of English manuscript verse of the seventeenth and eighteenth centuries: it was agreed at an early stage that this should be a Leeds-based tutorial, using exclusively Leeds manuscripts. We had the poems digitally photographed by the University Media Services, and found someone to make preliminary transcriptions. So far so good. But the work of establishing master transcriptions for the computer program to work from—a task that devolved mainly on Oliver Pickering—proved to be time-consuming, not least because there were a number of basic issues that had to be resolved in respect of the rules for transcription: it became clear from an early stage that some traditional scholarly conventions could not straightforwardly be transferred to the electronic medium.

Some decisions flowed naturally from our conviction that the user should not be required to employ anything in the way of special codes, fonts, or key-combinations (which would inhibit ease of use) in order to produce an acceptable transcription. The rules therefore say that accents and similar diacritics are not to be reproduced, that superscript letters should be typed on the line like normal letters, and that aspects of the layout of the text (indentation, say, or different sizes of script) should be ignored. They also lay down that archaic letter forms (for example, long s) are to be rendered by their modern equivalents, and that the expansion of contractions, suspensions, and other forms of abbreviation is to be indicated by enclosing the supplied letters within curly brackets, or braces. (Round brackets were ruled out for this purpose because they sometimes occur in manuscripts of this period.) A related special rule—because involving a letter form, not an abbreviation—is that the form "y" when used to signify the sound /th/ should be rendered as {th}, that is also within braces.

Another group of decisions concerned the treatment of scribal mistakes or changes of mind. Scribal insertions clearly have to be indicated, and here we decided that there had to be a way of showing precisely which portion of text had been introduced. The rule, in consequence, is that inserted text should be signalled by caret marks ^ ^ on either side of the insertion. This is obviously a departure both from what is written on the page and from normal scholarly practice, but we cannot see any objection to the procedure in the context of the package (and the point of the exercise, after all, is hardly to create another facsimile). Scribal corrections also need to be taken into account, and in this case we say that where the process of correction extends to the erasure of one form of words and the substitution of another, only the latter (that is, the scribe's apparent final intention) should be reproduced—that's to say, the student is not asked also to include the rejected form. The same applies to scribal deletion of words or letters without substitution: deleted material is not to be included in the transcription at all. In a critical edition there would of course be a case for indicating the fact of a deletion or substitution, but we don't see that this is necessary for the purposes of the tutorial. As for scribal errors—places where it seems probable that the scribe has copied inaccurately so that the text does not make sense—we are quite firm that there should be no attempt to emend, which we do not see as part of the exercise. This is not a package to help you learn to edit.

It was also necessary to decide what to do about elements of the manuscript where there was not necessarily one correct transcription. Punctuation is one example—where there is frequently ambiguity that is irresolvable without returning to the original (and not always then)—and capitalization is another. Andrew Booth overcame this problem by building different levels of tolerance into the program on a case-by-case basis, allowing the user to transcribe either a comma or a full stop, or either an upper-case or lower-case letter, at particular predetermined points. In the end we decided that it was better to say that punctuation should be ignored altogether (although transcriptions including it are not marked wrong), in the belief that it is more important for students to learn how to recognize letter forms than have to struggle with such minutiae (and perhaps become discouraged). However, the ability to allow either upper-case or lower-case, at individual points in the manuscripts where there is a real ambiguity in the originals themselves, has been a great advantage, which really came into its own during a second phase of development (funded in 2001–02 by the University's Communication and Information Technology in the Curriculum fund) that saw the addition of a further fifteen documents, this time of a broadly historical nature and stretching back into the sixteenth century (but all still in English). The rules for transcription devised for the original group of poems were usefully tested by the expansion of the tutorial in this way, and some modifications were consequently made to the method of handling abbreviated words.

II

A key aspect of the transcription program was that it should be deliverable via a WWW page. The first version of the program was written in Javascript, using Dynamic HTML (DHTML) and the Document Object Model (DOM). In this version the user is presented with, on the left, a window containing the manuscript to be transcribed and, on the right, an area in which the transcription can be entered. There is also a button which, when pressed, causes the student's transcription to be compared with a "master" version (Figure 12.1). The program then returns the transcription to the student with any errors highlighted in red. The checking procedure can be exact, in which case the student's transcription must exactly match the master.

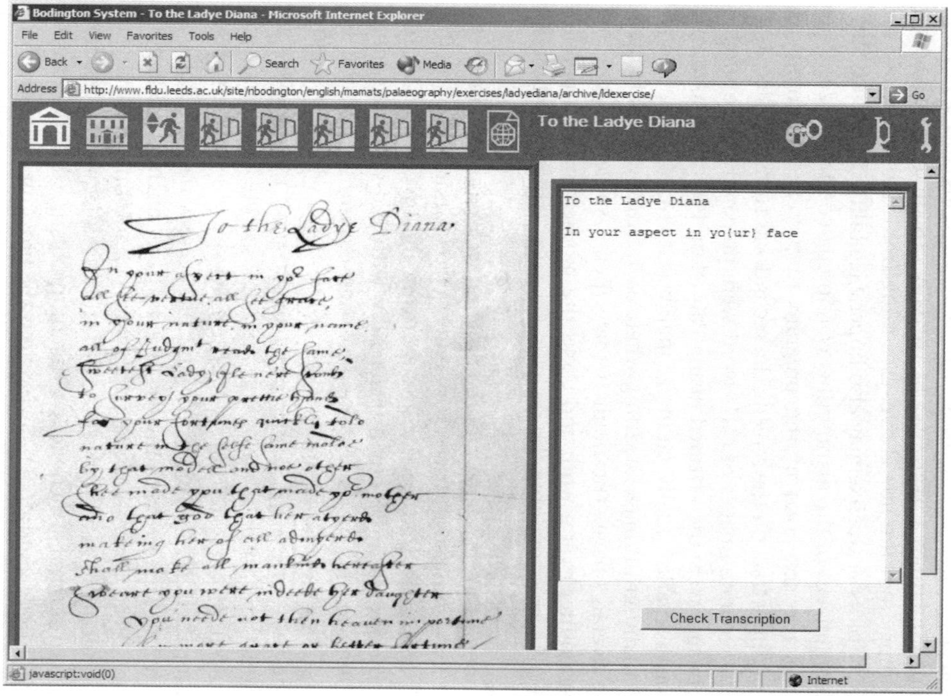

Figure 12.1: Screen shot 1

Alternatively, the checking can be looser in that the master can contain alternative spellings of individual words (or, as described above, renderings of individual letters), each of which is regarded as correct. A feature of many secretary hand manuscripts is their economical use of space. The lines tend to be close together and may overlap. Sometimes a word may be partially obscured by characters descending from the line above, ascending from the line below, or both. Ink blots or the like may also impede legibility. To address this problem, images of the manuscripts were prepared in two sizes. The first was the size used in the program to present the complete manuscript extract on the screen. The second was twice this size, and from it, separate images of each word were prepared. These single-word images were then treated (using Paintshop Pro) to remove any extraneous marks. Finally, an image map was created for each "complete" manuscript image, defining where each single word is located in the overall image. The result is that when a student moves the mouse pointer over a word and clicks the mouse button, a pop-up image appears containing the digitally "cleaned" single-word image (Figure 12.2).

Although this first version of the program worked reasonably well, it had some drawbacks. It was very much a work in progress, and the first students assisted greatly in suggesting improvements. At this stage the program did not interact fully with the virtual learning environment and partially completed transcriptions could not be saved. This was a significant problem for students who preferred to work in short bursts and wished to be able to return to complete an individual task without starting again from scratch. In addition no sample letter forms (or other "help" files) were provided as part of the program itself; in the initial trialling of the material, examples of letter forms were simply provided in hard copy at the introductory seminar. It was clearly essential to provide such information online to assist students as they experienced difficulties while working on their own. Furthermore, there was a major problem with the use of DHTML and Javascript/DOM, which put the program at the mercy of the web browser. Each browser seemed to implement the DOM incompletely and to interpret the DOM specification in different ways. As it started, the program had to identify the user's browser and then implement browser-specific code which had to be changed each time a new version of the browser became available. When funding

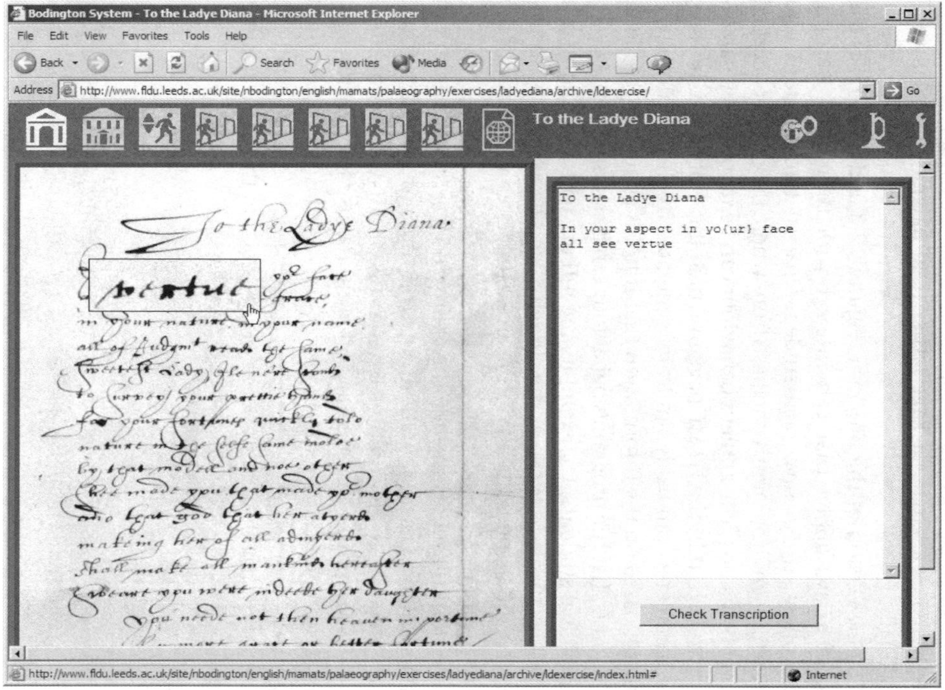

Figure 12.2: Screen shot 2

became available for the continuation of the project, in 2001, it was decided to abandon DHTML and rewrite the program from scratch in Java. By using the Sun Java plug-in, the program would be better insulated from the eccentricities of individual browsers. It would also load over the WWW much faster, since the many images needed can be delivered as a single package rather than having to be delivered individually.

The new version of the program (Figures 12.3 and 12.4) takes the form of a Java applet that can interact with the virtual learning environment. It can save each student's partial transcriptions into the learning environment's database from where they can be recalled on later occasions. The applet's behaviour can be modified by parameters passed to it in the HTML page in which it is embedded. In turn, these parameters can be dynamically generated by the learning environment, based on the information that it holds about each student. The result is that the program's behaviour can be tailored to the needs of individual students. For example, the full master transcription can be made available to those of the students who have made a reasonable attempt at the transcription exercise, but now want to see "the answers" and move on.

Other improvements include the ability to view the manuscript in portrait or landscape orientation; to scroll through the image if not all of it is visible; to view sample letter forms; and to read a contextual and palaeographical commentary on each manuscript. The rules for transcription, crucial to the student, are now also readable from within each exercise, as is the master transcription (if made available by the tutor); in the earlier version of the program these two features formed part of the tutorial's introductory pages.

III

The program has so far been used three times in the context of the MA in Renaissance English Literature for which it was originally intended, and by a small number of students beginning Ph.D. study in the School of English. Students have commented very positively upon it in feedback, and there is no doubt that they have found it enjoyable to use. The most successful teaching sessions have occurred when—after a short introductory demonstration—they have been able to try the program for themselves for the first time at PCs within

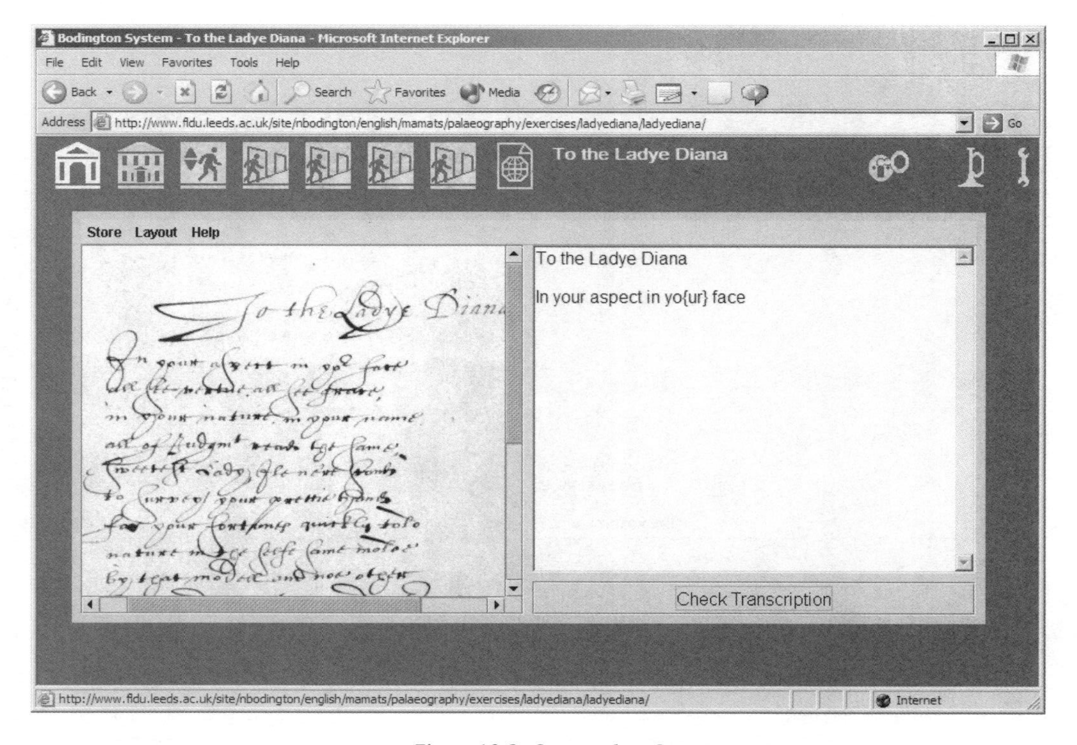

Figure 12.3: Screen shot 3

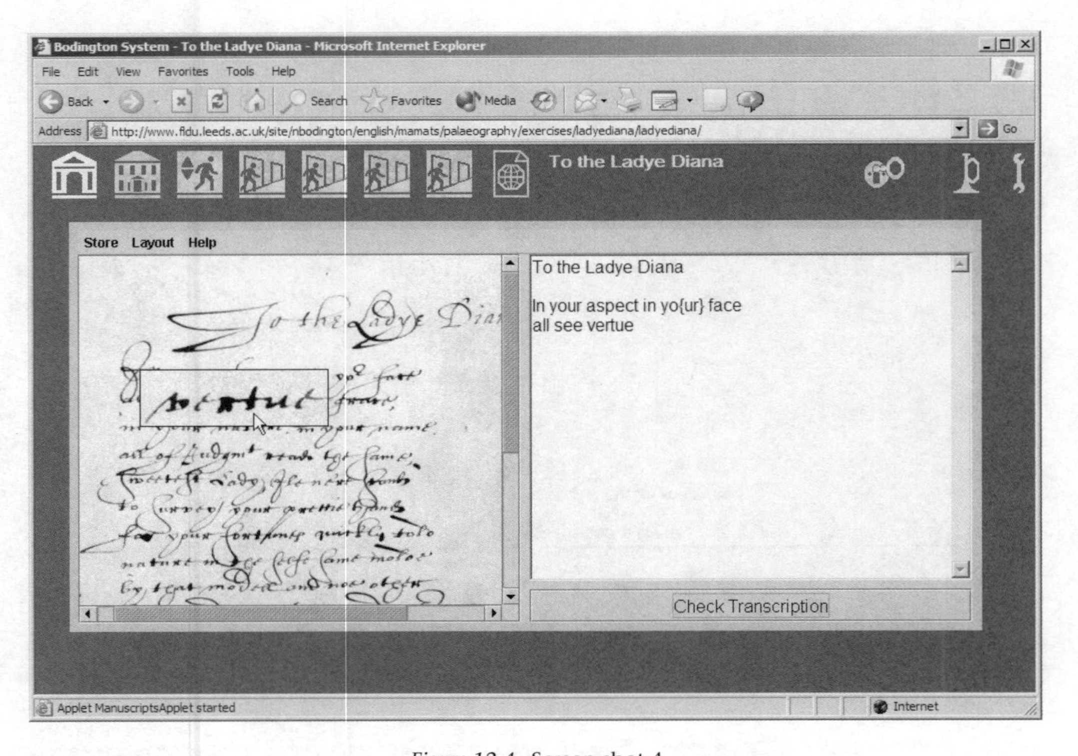

Figure 12.4: Screen shot 4

a dedicated teaching cluster, with one or more tutors circulating amongst them and giving assistance and / or encouragement as required. Even in these IT-saturated times, it is easy to underestimate the degree to which many students, particularly in arts disciplines, can still be intimidated by applications with which they are unfamiliar.

The authors of the tutorial are aware that, despite the existence of the pull-down sample letter forms and commentaries (which go so far as to explain how to transcribe some particularly difficult words), the deliberately simple screen arrangement of manuscript image on one side and blank space on the other can lead to an exclusively trial-and-error attitude to the exercises on the part of students: type, click the "check transcription" button, see which words are wrong, go back, try again, and so on. That is to say, it is apparent that some students appear either to like the challenge of puzzling it out for themselves, or else quickly forget about the existence of the various forms of "help" when faced with the image and empty screen. Nevertheless—and given the tutorial's deliberately limited aims in respect of the science of palaeography—we feel we now have the pedagogical balance about right. Learning how to read sixteenth- and seventeenth-century manuscripts should not be made too easy. Prompts should not be provided at every point. Some forms of assistance are provided, and it is up the student to remember to exploit them. We would say from experience, however, that the presence of tutors at the first group session, moving between students, is a vital part of the initial learning process. Others might, of course, wish to exploit the material differently. At Leeds we have not so far had follow-up sessions—partly because "palaeography" does not have its own module within the Renaissance MA—but have required the students to complete the transcription of a number of different manuscript extracts in their own time. Fortunately the results have generally been good.

IV

But work goes on. One development is that Oliver Pickering and Andrew Booth received further funding from the University in 2003—this time from the HEFCE-backed Teaching Quality Enhancement Fund—to develop a medieval version of the tutorial, again using exclusively Leeds University Library manuscripts. This work was carried out by a specially appointed research fellow, Geert De Wilde, and the new tutorial was brought into use in session 2003–04, principally within the MA in Medieval Studies although with the enthusiastic support of other relevant arts departments. Secondly, as a result of successful presentations at conferences, the developers are faced with a decision about whether—and how—to make the tutorial available for use outside the University of Leeds. Two possible models present themselves, both under active investigation at the time of writing: (1) Sale and distribution of the exercises on a CD, which should not be technically difficult because we have already written a Java application that can be used to replace the web browser and launch the transcription applet on a free-standing computer. We would expect the CD route (perhaps one CD for the "Renaissance" exercises and another for the medieval) to be particularly popular with people, both within and outside academic institutions, working independently. (2) Access to a parallel version of Bodington Common, restricted to the palaeography exercises alone and available via an annual subscription. This, we envisage, would be an appropriate model for postgraduate classes in other universities.

But making the tutorial available to others—and thus more free-standing—would almost certainly mean additional contextual work. Thus some broader introduction to the development of secretary hand (or of the range of medieval hands) would probably after all be necessary, as would more attention to divergent and difficult letter forms. And we would also have to decide—with thirty images in the Renaissance version alone—in which order, or in what sort of grouping, the different manuscripts should be presented, a decision that has so far been avoided. For example should we present them in order of difficulty, presumably working from easier to harder; or chronologically, so far as that can be determined; or by type of hand, or type of document? A redesigned front end for the tutorial would also be necessary, one that would enable students without the benefit of the first, tutor-guided, practice sessions to find their own way in to the package as a whole, and to plot their path through the various examples in a helpful fashion.

Meanwhile there is no doubt that the palaeography project—as it's known—has achieved its aims within the University of Leeds. It has made teaching and learning the rudiments of reading secretary hand (and now medieval hands) fresh and enjoyable, by creating a challenge for students that is simultaneously fun to attempt. There is also the immediacy of working on manuscripts known to be close at hand, and which in some cases have actually been shown and discussed in a preceding class (on manuscript verse miscellanies).

Appendix: Key Individuals

Randall Bass is Executive Director, Center for New Designs in Learning and Scholarship (CNDLS) and Assistant Provost for Teaching and Learning Initiatives at Georgetown University. He directs, among other projects, the "American Studies Crossroads Project," "T-AMLIT: Elecronic Archives for Teaching the American Literatures," and "The Heath Anthology of American Literature Newsletter *Online*."

Tim Berners-Lee is the author, with Mark Fischetti, of *Weaving the Web: the Original Design and Ultimate Destiny of the World Wide Web by its Inventor* (1999). He is generally credited with originating the concept of the WWW as a distributed hypertext system or linked information systems for the management of knowledge generated by large-scale experimental projects at CERN.

Vannevar Bush originated the concept of hypertext. In the article "As We May Think," published in *Atlantic Monthly* (1945), Bush describes a photo-electrical-mechanical device called a Memex, which could make and follow links between documents on microfiche. He proposed the development of an analogue computer, the Rockefeller Differential Analyser, but this idea became obsolete with the advance of digital computing.

Robert Cailliau recently retired from CERN where he had worked since 1974. In 1990, with Tim Berners-Lee, he proposed a hypertext system for access to CERN documentation. In 1995, the ACM attributed the Software System Award to Cailliau and Tim Berners-Lee for their work on the World Wide Web. He is the author, with James Gillies, of *How the Web was Born: the Story of the World Wide Web* (2000).

Douglas Engelbart, inventor of the computer mouse, developed Vannevar Bush's idea for hypertext. In the 1960s, he worked at the Stanford Research Institute where he and his colleagues, William K. English and John F. Rulifson, working in the Augmentation Research Centre, created the Online System (NLS), the world's first implementation of hypertext. NLS enabled hypertext linkage, word processing, e-mail, and teleconferencing within a full windowing software environment.

N. Katherine Hayles is Professor of English at UCLA and a pioneer in the field of literature and science. Her books include *The Cosmic Web: Scientific Field Models and Literary Strategies in the Twentieth Century* (1984), *Chaos Bound: Orderly Disorder in Contemporary Literature and Science* (1990), *How We Became Posthuman: Virtual Bodies in Cybernetics, Literature and Informatics* (1999) and *Writing Machines* (2002).

George Landow is Professor of English and Art History at Brown University. He is the author of *Hypertext: the Convergence of Contemporary Critical Theory and Technology* (1992, 2nd edn 1997) and *Hyper/Text/Theory* (1994). Web sites he has created and maintains are: *The Victorian Web, Postimperial and Postcolonial Literature in English* and *Cyberspace, Hypertext and Critical Theory.*

Jerome McGann is John Stewart Bryan Professor of English at the University of Virginia, with a specialism in the theory of textuality and media. He is the author of the influential book *Radiant Textuality: Literary Studies after the World Wide Web* (2001) and *The Complete Writings and Pictures of Dante Gabriel Rossetti. A Hypermedia Research Archive.*

Stuart Moulthrop is a practitioner and influential theoretician of hypertext fiction. His hyperbooks include *Dreamtime, The Garden of Forking Paths* and *Victory Garden.*

Janet Murray is Professor of Literature, Communication, and Culture at the Georgia Institute of Technology. She specializes in digital media curricula, interactive narrative, story/games, interactive television, and large-scale multimedia information spaces. Her book, *Hamlet on the Holodeck: the Future of Narrative in Cyberspace* (1997), pioneered the investigation of broadband art, information, and entertainment environments.

Theodor Holm (Ted) Nelson coined the word "hypertext" in 1965. He is the author of *Dream Machines* (1974), *Literary Machines* (1987) and *Hypertext and Hypermedia* (1990), and many articles on hypertext and knowledge management. Much of Nelson's effort has been devoted to developing an ambitious knowledge management system, Xanadu, which he calls "a universal instantaneous hypertext publishing network."

Works Cited

Abbate, Janet. *Inventing the Internet.* Cambridge, MA: MIT Press, 1999.

American Culture Studies Onweb. http://www.acs-onweb.de.

The American Memory Project. The Library of Congress. http://memory.loc.gov.

Aristotle. *Aristotle.* Trans. Philip Wheelwright. New York: Odyssey Press, 1951.

———. *The Rhetoric of Aristotle.* Trans. Lane Cooper. Englewood Cliffs, NJ: Prentice-Hall, 1960.

Arnold, Matthew. "The function of criticism at the present time" (1864). *Matthew Arnold: Selected Poetry and Prose.* Ed. Frederick L. Mulhauser. New York: Holt, Rinehart, 1953.

Asensi, Manuel. *J. Hillis Miller, or, Boustrophedonic Reading.* Trans. Mabel Richart. Palo Alto: Stanford University Press, 1999.

Ashcroft, Bill, Gareth Griffiths and Helen Tiffin, eds. *The Post-Colonial Studies Reader.* London: Routledge, 1995.

———. *Post-Colonial Studies: the Key Concepts.* London: Routledge, 2000.

Atwan, Robert. *Convergences.* New York: Bedford St. Martin's, 2002.

Barbasi, Alfred Laszlo. *Linked.* New York: Perseus, 2002.

Bass, Randy. *Engines of Inquiry: a Practical Guide for Using Technology to Teach American Culture.* Washington, DC: American Studies Crossroads Project, 1998.

Batschelet, Margaret. *Web Writing/Web Designing.* New York: Allyn and Bacon, 2001.

Baudrillard, Jean. *America.* Trans. Chris Turner. London: Verso, 1988.

Bligh, Donald. *What's the Use of Lectures?* Oxford: Oxford University Press, 1998.

Bolter, J. D. *Writing Space: the Computer, Hypertext and the History of Writing.* Hillsdale NJ: Lawrence Erlbaum and Associates, 1991.

Bolter, J. David and Richard Grusin. *Remediation: Understanding New Media.* Cambridge, MA: MIT Press, 1999.

Boyle, James. *Shamans, Software and Spleens: Law and the Construction of the Internet Society.* Cambridge, MA: Harvard University Press, 1996.

———. "The second enclosures movement." *Law and Contemporary Problems* 66 (2002): 33–74.

Boynton, R. S. "The Tyranny of Copyright?" *New York Times Magazine* 25 January 2004, 40–6.

Brennan, J. and R. Williams. *The English Degree and Graduate Careers. LTSN Report Series* 2 (January 2003).

Brown, Michael. *Who Owns Native Culture.* Cambridge, MA: Harvard University Press, 2003.

Burbules, Nicholas. "Rhetoric of the Web." *Page to Screen: Taking Literacy into the Electronic Era.* Ed. Illan Snyder. London: Routledge, 1998. 102–22.

Burke, Kenneth. "Literature as equipment for living". *Perspectives by Incongruity*. Ed. Stanley Edgar Hyman and Barbare Karmiller. Bloomington: Indiana University Press, 1964. 100–09.

Burke, Mary. "The travellers: Ireland's ethnic minority." *The Imperial Archive* 23 June 1999. http://www.qub.ac.uk/en/imperial/ireland/travellers.htm (accessed 3 January 2004).

Business Weekly. 17 November 2003.

Carnevale, D. "Study of Wisconsin professors find drawbacks to course-management systems." *Chronicle of Higher Education* 4 July 2003. http://chronicle.com/weekly/v49/i43/43a02602.htm (accessed 10 March 2004).

Carroll, Lewis. *The Annotated Alice*. Ed. Martin Gardner. New York: World Publishing Company, 1960.

Castells, Manuel. *The Information Age: Economy, Society and Culture*. Vol. 1. *The Rise of the Network Society*. Malden, MA: Blackwell, 1996.

Codd, E. F. *The Relational Model for Database Management, Version 2*. Reading, MA: Addison-Wesley, 1990.

Coombe, Rosemary. *The Cultural Life of Intellectual Properties: Authorship, Appropriation and the Law*. Durham: Duke University Press, 1998.

Condron, F., M. Fraser, and S. Sutherland. *Guide to Digital Resources for the Humanities*. Oxford: Oxford University Press, 2000.

Connexions. http://cnx.rice.edu.

Connors, Robert. "Textbooks and the evolution of the discipline." *Selected Essays of Robert J. Connors*. Ed. Lisa Ede and Andrea A. Lunsford. New York: Bedford St Martin's, 2003. 99–118.

Crane, Walter. *Of the Decorative Illustration of Books Old and New*. London: George Bell and Sons, 1972 [1896].

Daedalus Group. http://www.daedalus.com.

Daley, Elizabeth. "Expanding the concept of literacy." *Educause Review* (March/April 2003): 33–40.

Davidson, Cathy and David Theo Goldberg. "A manifesto for the humanities in a technological age." *Chronicle of Higher Education: the Chronicle Review* 50:23 (13 February 2004): B7.

Dawson, Giles E. and Laetitia Kennedy-Skipton. *Elizabethan Handwriting, 1500–1650*. London: Faber and Faber, 1968.

DeLillo, Don. *End Zone*. Boston: Houghton Mifflin, 1972.

de Man, Paul. "The rhetoric of temporality." *Blindness and Insight: Essays in the Rhetoric of Contemporary Criticism*. 2nd edn. rev. Minneapolis: University of Minnesota Press, 1983. 187–228.

———. "Autobiography as de-facement." *The Rhetoric of Romanticism*. New York: Columbia University Press, 1984a. 67–81.

———. "Shelley disfigured." *The Rhetoric of Romanticism*. New York: Columbia University Press, 1984b. 73–123.

Dertouzos, Michael L. *What Will Be: How the New World of Information Will Change Our Lives*. New York: Harper Collins, 1998.

Dirlik, Arif. *The Postcolonial Aura: Third World Criticism in the Age of Global Capitalism*. Boulder: Westview, 1997.

Doctorow, Cory. *Down and Out in the Magic Kingdom* (2003), http://www.craphound.com/down/.

Dunlop, Nicholas. "The fiction of Peter Carey: a bibliographic project." *The Imperial Archive* 13 May 1998. http://www.qub.ac.uk/en/imperial/austral/careybib.htm. Rpt. as "The fiction of Peter Carey: a bibliographical project." *Contemporary Postcolonial and Postimperial Literature in English*. http:// www.postcolonialweb.org/australia/carey/bibl/bibliography1.HTML (accessed 30 December 2003.

Ede, Lisa and Andrea Lunsford. *Singular Texts/Plural Authors: Perspectives on Collaborative Writing*. Carbondale, IL: Southern Illinois University Press, 1990.

Eilola, Johndan Johnson. *Designing Effective Websites: a Concise Guide*. New York: Houghton Mifflin, 2002.

Eisner, Will. *Comics and Sequential Art*. Tamarac, FL: Poorhouse Press, 1985.

Engle, Michael. "Evaluating web sites." 1996 [rev. 2004]. http://www.library.cornell.edu/olinuris/ref/research/webeval.HTML (accessed 18 May 2004).

English Benchmarking Statement. Gloucester: Quality Assurance Agency for Higher Education, 2000. http://www.qaa.ac.uk/crntwork/benchmark/english.pdf.

Faddan, Aidan. "History, language, and the post-colonial question in Brian Friel's translations." *The Imperial Archive* 12 May 1998. http://www.qub.ac.uk/en/imperial/ireland/friel1.htm (accessed 2 January 2004).

Faddan, Aidan and Andy Morrison. "An interview with Declan Kiberd." *The Imperial Archive* 21 January 1999. http://www.qub.ac.uk/en/imperial/ireland/kiberd.htm (accessed 2 January 2004).

Fallon, Tara. "Indian mutiny of 1857–1858: its causes and consequences." *The Imperial Archive* 3 June 1997. http://www.qub.ac.uk/en/imperial/india/mutiny.htm (accessed 22 January 2005).

Feather, John. *Publishing, Piracy and Politics: an Historical Study of Copyright in Britain*. London: Mansell Publishing, 1994.

Feenberg, A. "No frills in the virtual classroom." *Academe* (Sept–Oct 1999): 26–31.

Feldman, Martha S. and James G. March, "Information in organizations as signal and symbol." *Administrative Science Quarterly* 26 (1981):171–86.

Flindt, N. "e-learning—Theoriekonzepte und Praxiswirklichkeit mit Fallstudien von SAS Deutschland und der Karls-Ruprechts-Universität Heidelberg." Heidelberg: unpublished manuscript, 2003.

Foucault, Michel. "What is an Author?" *Language, Counter-Memory, Practice: Selected essays and interviews*. Ed. Donald F. Bouchard. Ithaca, NY: Cornell University Press, 1977. 113–38.

Galloway, Alexander R. *Protocol: How Control Exists after Decentralization*. Cambridge, MA: MIT Press, 2004.

Gates, Bill with Nathan Myhrvold and Peter Rinearson. *The Road Ahead*. Rev. edn. New York: Penguin, 1996.

Gibson, William. *Neuromancer*. New York: Ace Books, 1984.

Grossman, Wendy. *Net.wars*. New York: New York University Press, 1999. (Available for download at http://www.nyupress.org/netwars/).
Guillory, John. *Cultural Capital: the Problem of Literary Canon Formation*. Chicago: University of Chicago Press, 1993.
Hanrahan, Michael. *English and IT. LTSN Report Series* 5 (December 2002).
Hayden, C. *When Nature Goes Public: the Making and Unmaking of Bioprospecting*. Princeton: Princeton University Press, 2003.
Hebdige, Dick. *Subculture: the Meaning of Style*. London: Methuen, 1979.
Heim, Michael. *Electric Language: a Philosophical Study of Word Processing*. New Haven: Yale University Press, 1987.
HUMBUL Humanities Hub. University of Oxford, 2000–2005. http://www.humbul.ac.uk Internet for English site (Resource Discovery Network, 2000–2005), http://www.vts.rdn.ac.uk/tutorial/english/.
Jay, Gregory. "Strategies and challenges in high-tech teaching" 24 June 1999. http://www.uwm.edu/~gjay/dartmouth/ (accessed 7 February 2005).
Johns, A. *The Nature of the Book: Print and Knowledge in the Making*. Chicago: University of Chicago Press, 1998.
Joyce, Michael. *Of Two Minds: Hypertext Pedagogy and Poetics*. Ann Arbor: University of Michigan Press, 1995.
Kewes, Paulina, ed., *Plagiarism in Early Modern England*. New York: Palgrave Macmillan, 2003.
King, K. S. "Designing 21st century educational networlds: structuring electronic social spaces." *Electronic Collaborator: Learner-Centered Technologies for Literacy, Appenticeship, and Discourse*. Ed. C. J. Bonk and K. S. King. Mahwah, NJ, and London: Lawrence Erlbaum, 1998. 365–83.
Kolb, D. A. *Experiential Learning*. Englewood Cliffs, NJ: Prentice Hall, 1984.
Kirschenbaum, Mathew. "Hypertext." *Unspun: Key Concepts for Understanding the World Wide Web*. Ed. Thomas Swiss. New York: New York University Press, 2000. 120–37.
Landauer, Thomas K. *The Trouble with Computers: Usefulness, Usability, and Productivity*. Cambridge, MA: MIT Press, 1995.
Landow, George. "Rhetoric of hypermedia: some rules for authors." *Hypermedia and Literary Studies*. Ed. Paul Delany and George Landow. Cambridge, MA: MIT Press, 1991. 81–104.
———. *Hypertext: the Convergence of Contemporary Critical Theory and Technology*. Baltimore: Johns Hopkins University Press, 1992.
———. *Contemporary Postcolonial and Postimperial Literatures in English*. http://www.postcolonialweb.org/ (accessed 5 January 2004).
Laurillard, Diana. *Rethinking University Teaching*. 2nd edn. Routledge: London, 2002.
Lauter, Paul. *Canons and Context*. New York: Oxford University Press, 1991.
Laverty, C., Andy Leger, Denise Stockley, Mary McCollam, Stefan Sinclair, Donna Hamilton, and Christopher Knapper. "Enhancing the classroom experience with learning technology teams." *Educase Quarterly* 3 (2003):19–25.
Lawrence, John S. and Bernard Timberg. *Fair Use and Free Inquiry*. 2nd edn. Norwood, NJ.: Ablex Publishing, 1989.

LeClair, Tom. *In the Loop: Don DeLillo and the Systems Novel*. Urbana: Illinois University Press, 1988.

Lessig, Lawrence. *The Future of Ideas: the Fate of the Commons in a Connected World*. New York: Random House, 2001.

———. *Free Culture: How big media uses technology and the law to lock down culture and control creativity*. New York: Penguin, 2004. (http://www.free-culture.cc/freecontent/).

Levin, Doug, Sousan Arafeh, Amanda Lenhart, and Lee Rainie. "The digital disconnect: the widening gap between Internet-savvy students and their schools." Pew Internet and American Life Project, 14 August 2002. http://www.pewInternet.org/PPF/r/67/report_display.asp (accessed 27 February 2005).

Literature Online. ProQuest Information and Learning Company, 1996–2005. http://lion.chadwyck.co.uk/.

Litvack, Leon. *The Imperial Archive*. http://www.qub.ac.uk/en/imperial/imperial.htm (accessed 7 February 2005).

Liu, Alan. "Voice of the shuttle" 1994 (rev. 2001). http://vos.ucsb.edu.

———. *Palinurus: the Academy and the Corporation Teaching the Humanities in a Restructured World*. March 1998. http://palinurus.english.ucsb.edu (accessed 6 August 2004).

———. *Laws of Cool*. Chicago: University of Chicago Press, 2004.

Lowe, Charles and Terra Williams. "Moving to the public: weblogs in the writing classroom." *University of Minnesota Blog Collective: Into the Blogosphere*. http://blog.lib.umn.edu/blogosphere/moving_to_the_public.HTML (accessed 27 February 2005).

Lovink, Geert. *Uncanny Networks: Dialogues with the Virtual Intelligentsia*. Cambridge, MA: MIT Press, 2002.

Mangler, Simon, Matthias Mechler, Benjamin Reichert, Arne Spieker, and Florian Wildemann, "E-Learning der Universität Heidelberg." Baden-Baden: unpublished manuscript, 2003.

Manovich, Lev. *The Language of New Media*. Cambridge, MA: MIT Press, 2000.

Martin, Brendan. "Douglas Coupland's *Generation X: Tales for an Accelerated Culture*: an alternative voice." *The Imperial Archive* 18 May 1998. http://www.qub.ac.uk/en/imperial/canada/coupland.htm (accessed 12 December 2003).

McLeod, Kembrew. *Freedom of Expression®: Overzealous Copyright Bozos and Other Enemies of Creativity*. New York: Random House, 2005.

McGann, Jerome. *The Complete Writings and Pictures of Dante Gabriel Rossetti*. http://jefferson.village.virginia.edu/rossetti/.

———. *Radiant Textuality: Literature and the World Wide Web*. New York: Palgrave—now Palgrave Macmillan, 2001.

McLuhan, Marshall. *Understanding Media*. New York: Signet, 1964.

———. *The Medium is the Massage: an Inventory of Effects*. San Francisco: Hardwired, 1967 [rpt. 1996].

Merges, R., S. Menell, and M. A. Lemley. *Intellectual Property in the New Technological Age*. New York: Aspen Publishers, 2003.

Micklethwait, John and Adrian Wooldridge. *The Witch Doctors: Making Sense of the Management Gurus*. New York: Random House, 1996.

Miller, J. Hillis. *Black Holes*. Stanford: Stanford University Press, 1999.

Morrison, Andy. "The historical and colonial context of Brian Friel's *Translations*." *The Imperial Archive* 12 May 1998. http://www.qub.ac.uk/en/imperial/ireland/trans.htm (accessed 12 December 2003).

Negativland. *Fair Use: the Story of the Letter U and the Numeral 2*. Concord, CA: Seeland, 1995.

Newfield, Christopher. *Ivy and Industry: Business and the Making of the American University, 1880–1980*. Durham, NC: Duke University Press, 2003.

Northedge, A. and A. Lane. "Getting started." *The Sciences Good Study Guide*. Ed. A. Northedge, J. Thomas, A. Lane, and A. Peasgood. Milton Keynes: Open University, 1997. 1–23.

O'Loughlin, Jim. "Articulating Uncle Tom's Cabin." *New Literary History* 31.3 (Summer 2000):573–97.

———. " 'Grow'd again': articulation and the history of Topsy." Uncle Tom's Cabin *and American Culture Multimedia Archive* 2000. http://jefferson.village.virginia.edu/utc/interpret/exhibits/oloughlin/oloughlin.HTML.

Olsen, Florence. "Getting ready for a new generation of course-management systems." *Chronicle of Higher Education* 21 December 2001. http://chronicle.com/prm/weekly/v38/i17/17a02501.htm (accessed 10 March 2004).

Ong, Walter. "Writing is a technology that restructures thought." *Written Word: Literacy in Transition*. Ed. Gerd Baumann. Oxford: Oxford University Press, 1986. 23–50.

Page, Eimer. "Christophine site on the history and literature of the Caribbean." *The Imperial Archive* 27 April 2003. http://www.qub.ac.uk/en/imperial/carib/carib.htm (accessed 5 December 2003).

———. "Jean Rhys biography." *The Imperial Archive* 7 May 1997. http://www.qub.ac.uk/en/imperial/carib/rhysbio.htm (accessed 22 December 2003).

Palattella, John. "Formatting patrimony: the rhetoric of hypertext." *Afterimage* 23.1 (1995):13–21.

Palloff, R. M. and K. Pratt. *Lessons from the Cyberspace Classroom: the Realities of Online Teaching*. San Francisco, CA: Jossey-Bass Inc., 2001.

Patterson, L. R. *Copyright in Historical Perspective*. Nashville: Vanderbilt University Press, 1968.

Paulson, William. *Literary Culture in a World Transformed: a Future for the Humanities*. Ithaca: Cornell University Press, 2001.

Pew Internet and American Life Project, "Reports: online activities and pursuits" 29 February 2004. http://www.pewInternet.org/report_display.asp?r=113 (accessed 21 February 2005).

Plato. *The Collected Dialogues*. Ed. Edith Hamilton and Huntington Cairns. Princeton, NJ: Princeton University Press, 1961.

Preston, Jean F. and Laetitia Yeandle. *English Handwriting, 1400–1650*. Ashville: Pegasus Press, 1999.

ProQuest Learning. ProQuest Information and Learning Company, 2002–05. http://www.proquestlearning.co.uk/literature.

Readings, Bill. *The University in Ruin*. Cambridge, MA: Harvard University Press, 1996.

Reich, Robert B. *The Work of Nations: Preparing Ourselves for 21st-Century Capitalism*. New York: Random House, 1992.

Rheingold, Howard. *Tools for Thought*. Cambridge: MIT Press, 2000.

Richards, Thomas. *The Imperial Archive: Knowledge and the Fantasy of Empire*. London: Verso, 1993.

Rick, Jochen, Mark Guzdial, Karen Carroll, Lissa Holloway-Attaway, and Brandy Walker. "Collaborative learning at low cost: CoWeb use in English composition." Proceedings of CSCL 2002. http://newmedia.colorado.edu/cscl/93.pdf (accessed 27 February 2005).

Ricks, C. *Allusion to the Poets*. Oxford: Oxford University Press, 2002. 219–40.

Robins, Kevin and Frank Webster. *Times of the Technoculture: From the Information Society to the Virtual Life*. London: Routledge, 1999.

Rose, Mark. *Authors and Owners: the Invention of Copyright*. Cambridge, MA: Harvard University Press, 1993.

Rowe, John Carlos. *The New American Studies*. Minneapolis and London: University of Minnesota Press, 2002.

Saint-Amour, Paul. K. *The Copywrights: Intellectual Property and the Literary Imagination*. Ithaca, NY: Cornell University Press, 2003.

Schilling, Peter. "Bring in the geeks." *Educase Quarterly* 3 (2003a):12–13.

———. "From faculty to student: the evolution of educational technology." *Transformations: Liberal Arts in the Digital Age* 1:2 (2003b). http://www.colleges.org/transformations/index.php?q=node/view/40 (accessed 9 February 2004).

S-F Lovers. http://sflovers.org (accessed 17 August 2005).

Shumar, Wesley. *College for Sale: a Critique of the Commodification of Higher Education*. London: Falmer, 1997.

Slattery, Katharine. "Chinua Achebe and the language of the coloniser." *The Imperial Archive* 19 May 1998. http://www.qub.ac.uk/en/imperial/nigeria/language.htm (accessed 14 December 2003).

Snow, C. P. *The Two Cultures and the Scientific Revolution*. New York: Cambridge University Press, 1959.

Stephenson, Neal. *Snow Crash*. New York: Bantam Books, 1992.

Stewart, Nicholas. "Magic realism as post-colonialist device in *Midnight's Children*." *The Imperial Archive* 21 June 1999. http://www.qub.ac.uk/en/imperial/india/rushdie.htm (accessed 27 December 2003).

Stewart, Thomas A. *Intellectual Capital: the New Wealth of Organizations*. New York: Doubleday, 1997.

Strassmann, Paul A. *Information Payoff: the Transformation of Work in the Electronic Age*. New York: Macmillan—now Palgrave Macmillan, 1985.

Strunk, William and E. B. White. *Elements of Style*. New York: MacMillan, 1979.

Tufte, Edward. "PowerPoint is evil." *Wired Magazine* 11.09 (September 2003). http://www.wired.com/wired/archive/11.09/ppt2.HTML (accessed 18 May 18, 2004).

Ulmer, Gregory. *Heuretics*. Baltimore and London: Johns Hopkins University Press, 1994.

Unsworth, J. M. "The next wave: liberation technology." *Chronicle of Higher Education* 30 January 2004. http://chronicle.com/weekly/v50/i21/21b01601.htm (accessed 18 February 2004).

Vaidhyanathan, Siva. *Copyrights and Copywrongs: the Rise of Intellectual Property and how it Threatens Creativity*. New York: New York University Press, 2001.

———. *The Anarchist in the Library: How the Clash between Freedom and Control is Hacking the Real World and Crashing the System*. New York: Basic Books, 2004.

Victorian Web. http://www.victorianweb.org.

"Virtual seminars for teaching English literature" Project. University of Oxford, 1998. http://www.oucs.ox.ac.uk/ltg/projects/jtap.

Watkins, Evan. *Throwaways: Work Culture and Consumer Education*. Stanford: Stanford University Press, 1993.

Watts, Duncan J. *Six Degrees: the Science of a Connected Age*. New York: Norton, 2003.

Weibel, Peter and Timothy Druckery. *Net Condition: Art and Global Media*. Cambridge, MA: MIT Press, 2001.

Weller, Martin. *Delivering Learning on the Net*. London: Kogan Page, 2002.

Wideman, John Edgar. *Fever: Twelve Stories*. Harmondsworth: Penguin, 1990.

Williams, Jeffrey. "Brave new university." *College English* 61 (1999):742–51.

Williams, Raymond. *Marxism and Literature*. Oxford: Oxford University Press, 1977. 121–7.

Wishart, Adam and Regula Boschler. *Leaving Reality Behind: the Struggle for the Soul of the Internet*. London: Fourth Estate, 2002.

Woodmansee, Martha and Peter Jaszi, eds. *The Construction of Authorship: Textual Appropriation in Law and Literature*. Durham: Duke University Press, 1994.

Wyer, Conor. "The empire rides back." *The Imperial Archive* 13 May 2001. http://www.qub.ac.uk/en/imperial/transnational/Cycling.htm (accessed 3 December 2003).

Zuboff, Shoshana. *In the Age of the Smart Machine: the Future of Work and Power*. New York: Basic Books, 1988.

Further Reading

Akscyn, Robert M., Donald L. McCracken, and Elise A. Yoder. "KMS: a distributed hypermedia system for managing knowledge in organizations." *Communications of the ACM* 31.7 (July 1988):820–35.

Amiran, Eyal and Unsworth, John. "Postmodern culture: publishing in the electronic medium." *The Public-Access Computer Systems Review* 2.1 (1991):67–76.

Barrett, Edward, ed. *Text, ConText, and HyperText*. Cambridge: MIT Press, 1988.

———. *The Society of Text: Hypertext, Hypermedia, and the Social Construction of Information*. Cambridge: MIT Press, 1989.

Berk, Emily and Joseph Devlin, eds. *Hypertext/Hypermedia Handbook*. New York: McGraw-Hill, 1991.

Bolter, Jay David. *Writing Space: the Computer, Hypertext, and the History of Writing*. Hillsdale, NJ: Lawrence Erlbaum, 1991.

Burnett, Kathleen. "Toward a theory of hypertextual design." *Postmodern Culture* 3.2 (January 1993): n.p.

Bush, Vannevar. "As we may think." *Atlantic Monthly* July 1945: 101–08.

Conklin, E. Jeffery. "Hypertext: an Introduction and Survey." *IEEE Computer* 20 (September 1987):17–41.

Delany, Paul and George P. Landow, eds. *Hypermedia and Literary Studies*. Cambridge: MIT Press, 1991.

Ernst, Josef. "Computer poetry: an act of disinterested communication." *New Literary History* 23 (Spring 1992):451–65.

Handa, Carolyn, ed. *Computers and Community: Teaching Composition in the Twenty-First Century*. Portsmouth NH: Boynton/Cook, 1990.

Holdstein, Deborah H. and Cynthia L. Selfe, eds. *Computers and Writing: Theory, Research, Practice*. New York: The Modern Language Association of America, 1990.

Jonassen, David H. *Hypertext/Hypermedia*. Englewood Cliffs, NJ: Educational Technology Publishers, 1989.

Landow, George P. "The rhetoric of hypermedia: some rules for authors." *Journal of Computing in Higher Education* 1 (1989):39–64. Rpt. in Delany and Landow. 81–103.

———. *Hypertext: the Convergence of Contemporary Critical Theory and Technology*. London: Johns Hopkins University Press, 1992.

———. *Hyper/Text/Theory* London: Johns Hopkins University Press, 1994.

McLuhan, Marshall. *The Gutenberg Galaxy: the Making of Typographic Man*. Toronto: University of Toronto Press, 1962.

Nelson, T. H. *Hypertext and Hypermedia*. San Diego: Academic Press, 1990.

Nyce, James and Paul Kahn, eds. *From Memex to Hypertext: Vannevar Bush and the Mind's Machine*. San Diego: Academic Press, 1991.

Rada, R. *Hypertext: From Text to Expertext.* Maidenhead, UK: McGraw-Hill, 1991.

Yankelovich, Nicole, Bernard J. Haan, Norman K. Meyrowitz, and Stephen M. Drucker, "Intermedia: the concept and the construction of a seamless information environment." *IEEE Computer* 21 (January 1988):81–96.

For more information/resources on teaching English (both print and web-based) please go to the following link on the English Subject Centre web site: http://www.english.heacademy.ac.uk/explore/resources/teachlib/index.php.

Index